ALLE ZEIT WACH 1842

U. B. Sleytr P. Messner D. Pum M. Sára (Eds.)

Crystalline Bacterial Cell Surface Layers

With 86 Figures

Springer-Verlag
Berlin Heidelberg New York
London Paris Tokyo

Professor Dr. Uwe B. Sleytr

Dr. Paul Messner

Dr. Dietmar Pum

Dr. Margit Sára

Zentrum für Ultrastrukturforschung
der Universität für Bodenkultur
Gregor Mendel Str. 33
A-1180 Wien

Library of Congress Cataloging-in-Publication Data. Crystalline bacterial cell surface layers. Includes index. 1. Bacterial cell walls–Congresses. 2. Glycoproteins–Congresses. I. Sleytr, U.B. (Uwe Bernd) QR77.3.C79 1988 589.9'504875 88-15996

ISBN-13: 978-3-642-73539-4 e-ISBN-13: 978-3-642-73537-0

DOI: 10.1007/978-3-642-73537-0

Softcover reprint of the hardcover 1st edition 1988

2131/3130-543210

Preface

This volume contains the proceedings of the EMBO Workshop on crystalline bacterial cell surface layers (S-layers) which was held at the Universität für Bodenkultur, Vienna, August 31 to September 2, 1987. The workshop was generously co-sponsored by the Erwin Schrödinger Society for Microsciences and the Austrian Society for Biotechnology. It was the second meeting on this topic held in Vienna, following the first in August 1984.

The workshop provided a forum for a comprehensive current view on the structure, chemistry, biosynthesis, morphogenesis, function (including pathogenicity), genetics, evolutionary relationships, and biotechnological applications of crystalline arrays of macromolecules on prokaryotic cell surfaces.

S-layers are the simplest biological protein (glycoprotein) membranes developed during evolution. They are composed of a single molecular species endowed with the ability to assemble into regular arrays by an entropy-driven process.

As shown in the up to date compilation in the appendix of this volume, S-layers are found on members of nearly every taxonomical group of walled eubacteria and represent an almost universal feature of archaebacterial cell envelopes. Yet, surprisingly enough, up to now S-layers have received astonishingly little attention amongst micro-biologists. Like other bacterial cell surface components, S-layers have evolved as a consequence of specific interactions between the cells and their environment. Since prokaryotes carrying S-layers are ubiquitous in the biosphere, it is now quite evident that the supramolecular concept of a 'porous crystalline surface layer' can provide organisms with an advantage of selection in quite diverse habitats. Up to now S-layers have been shown to function as protective coats, molecular sieves, and molecule and ion traps, as promoters for cell adhesion and surface recognition and frameworks, determining and maintaining cell shape or envelope rigidity. It seems also quite likely that S-layer-like membranes could have fulfilled barrier and supporting functions as required by self-reproducing systems (progenotes) during the early periods of biological evolution. S-layers are now generally part of complex supramolecular envelope structures and may function only through specific interactions with other envelope components. In this context the exact definition of what we understand as an "S-layer" seems a more difficult task than generally appreciated. Crystalline arrays of proteins are also detected as components of the outer membrane of Gram-negative eubacteria. Since these assemblies apparently require a lipid bilayer for maintaining their structural integrity, the suggestion was made at this workshop not to include crystalline outer membrane proteins within the concept of S-layers. Thus at the present level of knowledge, S-layers could be defined as "two-dimensional crystalline arrays of proteinaceous subunits forming surface layers on prokaryotic cells."

Further studies will determine which structural, morphogenetic, functional, and evolutionary relationships exist amongst proteins arranged in crystalline layers within or on the external surface of the walls of prokaryotic cells.

Finally it should be stressed that with our increasing knowledge due to fundamental studies on S-layers, a considerable potential for various biotechnological applications becomes evident. Identifications of sequence homology of the S-layer subunits of pathogenic species could lead to the design of peptide subunit vaccines. More detailed information in the biosynthesis, regulation of synthesis and secretion of S-layer protein should be of relevance for improving the transfer of recombinant DNA products out of the bacterial cell. Permeability studies on S-layers have shown that they can be used for the production of isoporous ultrafiltration membranes with a broad potential for diversification.

We hope that these proceedings will fulfill various purposes, such as serving as an introduction to the field of S-layer research for young scientists, as encouragement to enter this subject, and as a record and reference book on the present state of this field.

We would like to mention our appreciation for the organizational and secreterial assistance of Thomas Blahous, Waltraud Haase, Helene Hendling, Claudia Hurban, Klaus Jaeger, Andrea Scheberl and Almud Urban and special thanks to Helene Hendling for retyping the manuscripts.

Vienna, January 1988

Uwe B. Sleytr
Paul Messner
Dietmar Pum
Margit Sára

Contents

Contributors

You will find the addresses at the beginning of the respective contribution

Comparative Chemistry of the Rigid Cell Wall Component and its Phylogenetic Implications

O. Kandler

Botanisches Institut, Universität München
Menzingerstr. 67
D-8000 München 19
FRG

Introduction

Cell walls[1] were the first structural cell elements observed in plant tissue (Hook 1665) and even gave rise to the term "cell" (from the latin cellula = chamber). When it became evident that only the cells of plants and fungi, but not those of animals, possess a rigid wall consisting of distinct polymers, namely cellulose and chitin, respectively, this property was considered as one of the important cytological criteria separating the animal and plant kingdoms. The occurrence of cell walls in bacteria, first demonstrated by plasmolysis and dyeing techniques, was an important argument for including these organisms in the plant rather than the animal kingdom.

Eubacterial Cell Walls

It was also a cell wall characteristic which, at the end of the 19th century, led to the division of the bacteria in two main groups, the Gram-positive and the Gram-negative. However, the structural details and the chemistry of the bacterial cell wall remained a mystery until the second half of the 20th century. Only after methods for ultra-thin sectioning of bacteria had been developed, could the various types of cell envelope profiles be elucidated by electron microscopy. At about the

[1]In this paper, the term cell wall will be restricted to mean a rigid layer forming a sacculus enclosing an individual cell, which, after isolation, still exhibits the shape of the cell and is insensitive to detergents. Arrays of protein or glycoprotein subunits covering the surface of the cells will be referred to as surface layers (S-layers) (Sleytr and Glauert 1982). Layers of distinct mechanical stability extending over several cells will be termed sheats. Mucous substances surrounding single cells or groups of cells will be termed capsules. Irrespective of the chemical and physical nature of the layers surrounding the cells, their sum will be referred to as the cell envelope.

same time, the pioneering work of Salton (1964) made isolated cell walls available for chemical analysis. Although crude cell wall preparations contained several accessory polymers (teichoic acid, teichuronic acid, heteropolysaccharides etc.) the purified rigid sacculi, exhibiting the shape of the cells, consisted of only <u>one</u> particular polymer, murein (peptidoglycan; Weidel and Pelzer 1964). It is made up of a glucan component - alternating β(1,4)-glycosidically linked N-acetylglucosamine and N-acetylmuramic acid residues - and short peptides attached to the carboxyl group of muramic acid (Fig.1). Adjacent peptides are crosslinked by a peptide bond between the carboxyl group of the C-terminal

Fig.1. Comparison of the chemical structures of murein and pseudomurein

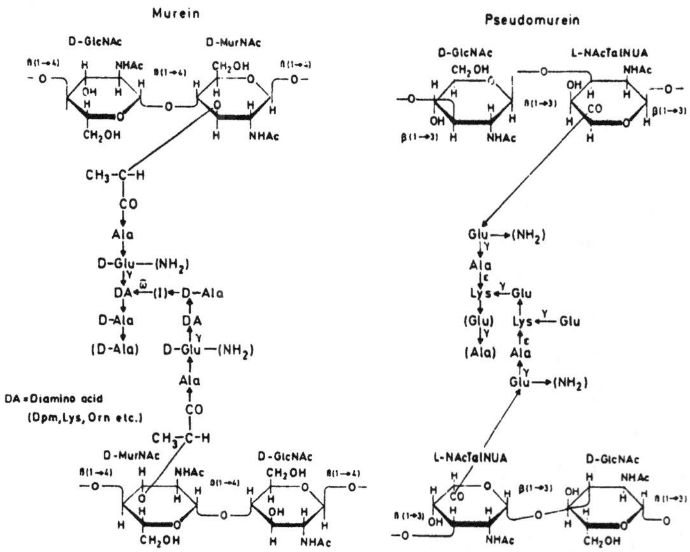

Fig.2. Schematic presentation of modifications of murein. Designation of mureintypes according to Schleifer and Kandler (1972)

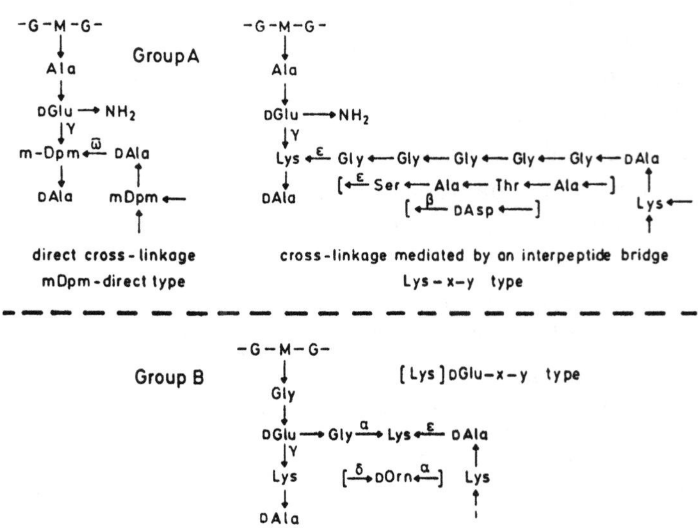

Table 1. Distribution of murein chemotypes among the eubacterial phyla. The phyla are based on 16 S rRNA (Woese 1987)

Eubacterial Phyla	Murein sacculus Width	Chemotype
Gram-positive	> 10nm	Variable
Radio-resistant cocci	> 10nm	Orn-Gly_2
Cyanobacteria	> 10nm	mDpm-direct
Prochloron	> 10nm	mDpm-direct
Spirochetes	< 10nm	Orn-direct
Green non-sulphur bacteria	< 10nm?	Orn-direct (?)
Purple bacteria	< 10nm	mDpm-direct
Green sulphur bacteria	< 10nm	"
Bacterioides	< 10nm	"
Thermotoga	patches ?	Lys-direct (?)
Planctomyces	No murein	
Chlamydia	No murein	

D-alanine and the N-terminus of the diamino acid, in many cases meso-diaminopimelic acid (mDpm), thus forming a huge macromolecule encompassing the entire bacterial cell.

Murein, in various chemotypes, is found in all eubacteria that possess a rigid sacculus (Table 1). Most phyla have a chemotype with mDpm as diamino acid and a direct cross-linkage between adjacent peptides (Fig.2). Also the murein of the spirochetes is directly crosslinked; however, mDpm is replaced by ornithine. Only the Gram-positive and the radiation-resistant cocci show a marked variation not only with respect to the kind of diamino acid (mDpm may be replaced with lysine, ornithine, diaminobutyric acid etc.), but also the mode of crosslinking. In the chemotypes of group A, crosslinking may be mediated by an interpeptide bridge consisting of only one amino acid residue or a short peptide (up to pentamers). The mode of crosslinking is even more markedly varied in the chemotypes of group B, which is only found in some taxa of coryneform bacteria. Here, crosslinking occurs between the α-carboxylgroup of D-glutamic acid and the C-terminal D-alanine of an adjacent peptide moiety via a diamino acid. Many variations of groups A and B - more than 100 are known at present - are more or less specific for certain genera or even higher taxa of the Gram-positives. However, in very progressive, presently fast-evolving groups of bacteria (e.g. micrococci, coryneformes etc.), even closely related species or subspecies may exhibit different chemotypes (Schleifer and Kandler 1972, Schleifer and Stackebrandt 1983).

The striking difference between the Gram-positives and the Gram-negatives in the extent of variation of the murein chemotypes may be due to the different molecular architecture of the cell walls in these two groups. While changes in the chemical structure may destroy the mono- or oligomolecular layers of the murein sacculi of Gram-negatives, the very thick multimolecular sacculi of Gram-positives obviously tolerate such changes.

Occasionally, the genetic information for the synthesis of murein may be lost during evolution giving rise to new lineages of murein-free bacteria. Such groups exhibit either no cell envelope, as is the case in *Mycoplasma*, a fairly recent offspring of Gram-positives (Woese 1987), or they exhibit a murein-free envelope which contains components resembling those of Gram-negatives, e.g. *Chlamydia* and *Planctomyces*. All three murein-free groups show a much faster evolution rate than the other bacteria (Woese 1987).

Archaebacterial Cell Walls

In contrast to eubacteria, no common cell wall polymer is found in archaebacteria (Kandler and König 1985, Kandler 1983). Only three taxa of archaebacteria stain Gram-positive and possess a rigid cell wall sacculus. However, the sacculus polymer is different in each of the three taxa.

The chemical structure of the sacculus of Methanobacteriales resembles murein in that it is also made up of a glucan and a peptide moiety. However, muramic acid is replaced by talosaminuronic acid, not known in any other organism, the glycosidic bonds are alternately β(1,4) and β(1,3), the peptide moiety contains no D-amino acids and the sequence of the peptide is basically different from that of murein (Fig.1). Thus, the polymer was called pseudomurein in order to indicate both the overall resemblance of the murein structure and the basic differences in details (Kandler and König 1985).

The cell walls of *Methanosarcina* consist of an acid polysaccharide, the so-called methanochondroitin composed of N-acetylgalactosamine and glucuronic acid (Kreisl and Kandler 1986; Fig.3). It resembles chondroitin, common to the connective tissue of animals. However, it is not sulphated

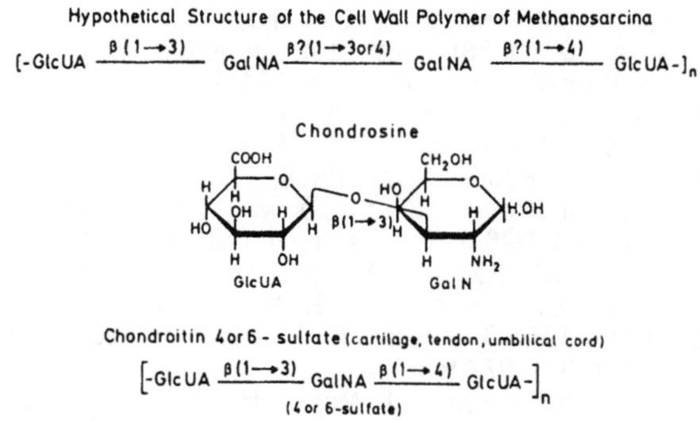

Fig.3. Comparison of the chemical structure of methanochondroitin in Methanosarcina with chondroitin in eukaryotes

and the repeating unit of the polymer is not a dimer, but a trimer containing 2 mol of N-acetylgalactosamine per mol of glucuronic acid.

The very thick cell walls of *Halococcus* are formed by a complex acidic heteropolysaccharide. The acid hydrolysate of the polymer contains several neutral and amino sugars, uronic acids, the very rare gulosaminuronic acid and glycin. Detailed studies have shown that N-acetyl-D-glucosamine is partially replaced by the unusual N-glycyl-glucosamine and that all neutral and amino sugar and uronic acid residues are highly sulphated. So far, the structure of this unusual polymer has only been partially elucidated (Schleifer et al. 1982).

With the exception of the cell envelope-less *Thermoplasma*, all other archaebacteria possess proteinacous cell envelopes, mostly S-layers, but in some cases, e.g. *Methanospirillum* and *Methanothrix*, they exhibit very complex structures.

Final Remarks

In conclusion, the distinct morphological and chemical diversity of the cell envelopes and cell wall polymers of the archaebacteria indicates that the common archaebacterial ancestor did not possess a specific cell wall or envelope polymer. Thus, in contrast to the situation found in the eubacterial kingdom, where at least murein is a common, although often modified heritage, the cell envelope components of archaebacteria have been envolved independently within the different lineages at a later state of evolution, a situation, also found in the eukaryotic kingdom.

References

Hook R (1665) Micrographia, London
Kandler O (1983) Cell wall structures and their phylogenetic implications. Zentralbl Bakteriol Hyg I Abt Orig C3:149-160
Kandler O, König H (1985) Cell envelopes of archaebacteria. In: Woese CR, Wolfe RS (eds) The bacteria, vol VIII. Archaebacteria, pp 413-457
Kreisl P, Kandler O (1986) Chemical structure of the cell wall polymer of Methanosarcina. Syst Appl Microbiol 7:293-299
Salton MRJ (1964) The bacterial cell wall. Elsevier, New York
Schleifer KH, Kandler O (1972) Peptidoglycan types of bacterial cell walls and their taxonomic implications. Bacteriol Rev 36:407-477
Schleifer KH, Stackebrandt E (1983) Molecular systematics of prokaryotes. Annu Rev Microbiol 377:143-187
Schleifer KH, Steber J, Mayer H (1982) Chemical composition and structure of the cell wall of *Halococcus morrhuae*. Zentralbl Bakteriol Hyg I Abt Orig C3:171-178
Sleytr UB, Glauert AM (1982) Bacterial cell walls and membranes. In: Harris JR (ed) Electron microscopy of proteins, vol 3. Academic Press, London New York, pp 41-76
Weidel W, Pelzer H (1964) Bagshaped macromolecules: a new outlook on bacterial cell walls. Adv Enzymol 26:193-232
Woese CR (1987) Bacterial evolution. Microbiol Rev 51:221-271

Comparative Aspects on Archaeobacterial Proteinaceous Cell Envelopes

H. König

Abteilung Angewandte Mikrobiologie, Universität Ulm
D-7900 Ulm
FRG

Archaeobacteria represent a third urkingdom of life (Woese 1987). It consists of heterogeneous organisms. Significant differences from the typical bacterial cell walls were already found in the 1950's (Houwink 1956). Common to all archaeobacteria is the lack of muramic acid, which proved to be true for some eubacterial genera. Unlike most eubacteria, the archaeobacteria possess no common cell wall polymer. They have developed diverse cell envelope types, which are restricted to the archaeobacteria (Fig.1; Kandler and König 1985; Sleytr et al. 1986; Baumeister and Engelhardt 1987). Because of their unusual cell envelope composition, the archaeobacteria exhibit a high degree of resistance against cell wall-specific antibiotics and lytic agents (Bock and Kandler 1985).

The cell walls of the Gram-positive archaeobacteria consist of pseudomurein, methanochondroitin or heteropolysaccharide. A double layered cell envelope composed of peptidoglycan and a proteinaceous crystalline surface layer (S-layer) is found only in the extremely thermophilic pseudomurein-containing genus *Methanothermus*. *Methanothermus fervidus* possesses a mannose- and 3-O-methylglucose-rich S-layer glycoprotein with a high amount of acidic amino acids (32 mol%). Some species of *Methanosarcina* also possess two cell envelope layers: an S-layer adherent to the outside of the cytoplasmic membrane and an outer methanochondroitin matrix.

Archaeobacteria with other types of cell envelopes react as Gram-negative. The typical cell wall profile of the Gram-negative eubacteria is absent. The most common archaeobacterial cell envelope is represented by a single surface layer (S-layer) composed of protein or glycoprotein subunits. Single S-layers are found in many organisms of the methanogenic branch and, despite their extreme biotopes, in all archaeobacteria of the extremely thermo(acido-)philic branch. The S-layers are often easily disintegrated and the mechanical and chemical stability does not parallel the extreme living conditions. In some cases proteinaceous surface layers exhibit remarkable stability. Examples are *Thermoproteus, Thermofilum, Methanospirillum* and *Methanothrix.* The cell wall-less *Thermoplasma* demonstrates that distinct cell envelopes are not required for life in extreme environments. The stabilization of the cytoplasmic membrane is most likely maintained by a mannose-rich glycoprotein and a lipoglycan.

The S-layer glycoprotein of *Halobacterium salinarium* was the first glycosylated protein found in prokaryotes (Mescher and Strominger 1976). The primary structure of the peptide and carbohydrate moiety has been elucidated (Wieland et al. 1982, Lechner and Sumper 1987). The carbohydrate moiety is composed of three different glycopeptides with novel linkage types and dolichyl derivatives serve as lipid carriers. Like the methanochondroitin, the sulphated halobacterial S-layer glycoprotein is reminiscent of components of the eukaryotic connective tissue, in particular the proteoglycans.

The occurrence of glycoprotein S-layers is not a common feature of all archaeobacteria. The surface layers of the methanococci are composed of nonglycosylated proteins (Nußer and König 1987).

In the case of *Methanospirillum* and *Methanothrix,* the cell envelope is more complex. Individual cells are surrounded by a most probably proteinaceous layer and several cells are enclosed in a filamentous sheath consisting of hoop-like rings. Several cells are separated by plugs. The sheath is mainly composed of protein and carbohydrate. Two structural models have been proposed. According to Patel et al. (1986), the sheaths of *Methanothrix cincilii* and *Methanospirillum hungatei* have an identical subunit arrangement with a p2 symmetry. A p1 symmetry of the subunits was proposed by Shaw et al. (1985).

The two- and three-dimensional arrangement of the S-layer proteins of the extreme thermophiles *Sulfolobus acidocaldarius* (Michel et al. 1980, Deatherage et al. 1983, Baumeister and Engelhardt 1987) and *Thermoproteus tenax* and *T. neutrophilus* (Messner et al. 1986, Wildhaber and Baumeister 1987) have been elucidated. In both organisms the outer surface of the S-layer appears smooth, while the inner side is sculptured. The structural analyses revealed, that the S-layer proteins are complex molecules forming different domains. In the case of

Sulfolobus acidocaldarius only 30% of the S-layer is occupied by protein, which gives rise to a rather spongy structure.

Archaeobacteria with different cell envelopes exhibit distinct modes of cell division. The Gram positive archaeobacteria multiply by septum formation, most Gram negatives with single S-layers by constriction. True branching or budding also occurs. Division of the sheathed cells of *Methanospirillum* and *Methanothrix* proceeds by two different modes. *Methanothrix* divides by the ingrowth of the spacer plugs from the sheaths followed by fragmentation of the sheath. Whereas *Methanospirillum* divides through wall septation followed by cell spacer ingrowth and filament splitting.

S-layers may be the earliest archaeobacterial cell envelope type, since only some genera of the methanogenic branch have developed other cell wall polymers. They form additional layers or replace the S-layers. Most likely the nonglycosylated surface layer proteins represent primitive cell envelope structures, because the information for the self-assembly is exclusively preserved in the amino acid composition. No additional activated compounds are required for the biosynthesis of the cell envelopes.

Profile	Representative Genera
●●●●●●●●●●● SL ═══════ CM	Halobacterium, Methanococcus, Thermoproteus, Sulfolobus, Pyrodictium
●● ●● ●● ●● PS ●●●●●●●●●●● SL ═══════ CM	Methanospirillum, Methanothrix
●●●●●●●●●●● CM	Thermoplasma
─────── MC ●●●●●●●●●●● SL ═══════ CM	Methanosarcina
─────── PM, HP ═══════ CM	Methanobacterium, Halococcus
●●●●●●●●●●● SL ─────── PM ═══════ CM	Methanothermus

Fig.1. Schematic profiles of archaeobacterial cell envelopes. *CM* cytoplasmic membrane; *HP* heteropolysaccharide; *PM* pseudomurein; *PS* protein sheath; *SL* proteinaceous surface layer (S-layer); in the case of *Methanospirillum* and *Methanothrix* the crystalline proteinaceous nature has to be proved. *MC* methanochondroitin. The cell envelope of *Methanosarcina* can also consist only of one layer

The S-layer subunits are diverse in respect to molecular weight, chemical composition, stability and arrangement. The microheterogenicity of the archaeobacterial S-layers is not only a suitable taxonomic marker for distinguishing morphologically and physiologically similar organisms but may be helpful to learn more about the molecular strategy and phylogeny of S-layers.

References

Baumeister W, Engelhardt H (1987) Three dimensional structure of bacterial surface layers. In: Harris JR, Horne RW (eds) Electron microscopy of proteins, vol 6. Academic Press, London New York, pp 109-154

Bock A, Kandler O (1985) Antibiotic sensitivity of archaebacteria. In: Woese CR, Wolfe RS (eds) The bacteria, vol 8. Academic Press, London New York, pp 525-544

Deatherage JF, Taylor KA, Amos LA (1983) Three dimensional arrangement of cell wall protein of *Sulfolobus acidocaldarius.* J Mol Biol 167:823-852

Houwink AL (1956) Flagella, gas vacuoles and cell wall structure in *Halobacterium halobium;* an electron microscope study. J Gen Microbiol 15:146-150

Kandler O, König H (1985) Cell envelopes of archaebacteria. In: Woese CR, Wolfe RS (eds) The bacteria , vol VIII. Academic Press, London New York, pp 413-457

Lechner J, Sumper M (1987) The primary structure of a procaryotic glycoprotein. Cloning and sequencing of the cell surface glycoprotein gene of halobacteria. J Biol Chem 262:9724-2929

Mescher MF, Strominger JL (1976) Purification and characterization of a procaryotic glycoprotein from cell envelope of *Halobacterium salinarium.* J Biol Chem 251: 2005-2014

Messner P, Pum D, Sára M, Stetter KO, Sleytr UB (1986) Ultrastructure of the cell envelope of the archaebacteria *Thermoproteus tenax* and *Thermoproteus neutrophilus.* J Bacteriol 166:1046-1054

Michel H, Neugebauer DC, Oesterhelt D (1980) The 2-d crystalline cell wall of *Sulfolobus acidocaldarius:* Structure, solubilization, and reassembly. In: Baumeister W, Vogell W (eds) Electron microscopy at molecular dimensions. Springer, Berlin Heidelberg New York, pp 27-35

Nußer E, König H (1987) Cell envelope studies on 3 species of *Methanococcus* living at different temperatures. Can J Microbiol 33:256-261

Patel GB, Sprott GD, Humphrey RW, Beveridge TJ (1986) Comparative analyses of the sheath structures of *Methanothrix concilii* GP6 and *Methanospirillum hungatei* GP1 and JF1. Can J Microbiol 32:623-631

Shaw PJ, Hills GJ, Henwood JA, Harris JE, Archer DB (1985) Three dimensional architecture of the cell sheath and septa of *Methanospirillum hungatei.* J Bacteriol 161:750-757

Sleytr UB, Messner P, Sára M, Pum D (1986) Crystalline envelope layers in archaebacteria. Syst Appl Microbiol 7:310-313

Wieland F, Lechner J, Sumper M (1982) The cell wall glycoprotein of halobacteria: Structural, functional and biosynthetic aspects. Zentralbl Bakteriol Hyg I Abt Orig C3:161-170

Wildhaber J, Baumeister W (1987) The cell envelope of *Thermoproteus tenax:* three-dimensional structure of the surface layer and its role in shape maintenance. EMBO J 6:1475-1480

Woese CR (1987) Bacterial evolution. Microbiol Rev 51:221-271

Evidence for the Glycoprotein Nature of Eubacterial S-Layers

P. Messner and U.B. Sleytr

Zentrum für Ultrastrukturforschung
Universität für Bodenkultur
Gregor Mendel-Str. 33
A-1180 Wien
Austria

So far, about 200 different species of archaebacteria and eubacteria from most phylogenetic branches have been identified to carry regularly arranged surface layers (S-layers) (Sleytr and Messner, 1983, Sleytr et al. 1986a). Thus, S-layers appear to be a universal rather than a unique feature of prokaryotic cells. Quantitative calculations have revealed that S-layers can account for up to 10% of the total cell protein and therefore represent the major cell protein.

About ten years ago Mescher and Strominger (1976) described the first prokaryotic glycoprotein, which was the S-layer glycoprotein of the archaebacterium *Halobacterium salinarium.* In the meantime, covalently linked carbohydrate chains have been observed on several other archaebacteria including the methanogens and the members of the thermophilic-acidophilic branch (Kandler 1982, Kandler and König 1985, König and Stetter 1986). In many archaebacteria such as the extreme thermophilic-acidophilic archaebacteria, some methanogens and some halobacteria, an S-layer composed of glycoprotein subunits represents the only wall component outside of the plasma membrane (Kandler and König 1985). The most detailed structural analyses on S-layer glycoproteins have been performed by Wieland and coworkers on *Halobacterium halobium.* Both O- and N-glycosidically linked glycan chains and a high content of sulphate have been found together with different uronic acids including the rare 3-O-methylgalacturonic acid or iduronic acid (Wieland et al. 1983, Lechner et al. 1985, Paul et al. 1986, Wieland et al. 1986). Dolichols are involved in the biosynthesis of these glycoproteins (Sumper 1987, Wieland 1988). As a conclusion, the overall structure of the halobacterial cell surface glycoprotein is reminiscent of eukaryotic proteoglycan-collagen complexes which clearly demonstrates the general relationship between the archaebacterial glycoproteins and those of eukaryotic organisms (Wieland 1988).

While the existence of archaebacterial glycoproteins is generally accepted, only a few investigations reported on the presence of eubacterial glycoproteins since the first description on thermophilic *Clostridia* (Sleytr and Thorne 1976). Most of the investigations were based on SDS-PAGE and subsequent periodic acid-Schiff (PAS)-staining of the gels. If any further chemical analyses were performed, the carbohydrates were generally detected by colorimetric assays (Table 1).

In our laboratory, S-layers of several thermophilic Bacillaceae were screened for the presence of glycoproteins. The data on the chemical compositions were obtained from colorimetric tests and GC- and amino acid analyses of highly purified S-layers from those strains which were initially identified by PAS-staining of SDS-polyacrylamide gels (Sleytr and Thorne 1976, Sleytr et al. 1986b).

To obtain evidence for the existence of eubacterial glycoproteins those strains were selected which revealed a positive PAS-staining behaviour on SDS-polyacrylamide gels and high sugar contents. Eventually, *Bacillus stearothermophilus* NRS 2004/3a with an oblique S-layer lattice and a sugar content of about 5% and *Clostridium thermohydrosulfuricum* L111-69 with a hexagonal S-layer lattice and ca. 8% carbohydrate were chosen for the further structural investigations.

After pronase digestion of the isolated S-layer glycoprotein of *Bacillus stearothermophilus* NRS 2004/3a and purification by gel permeation chromatography and ion exchange chromatography, two glycopeptides, designated GP I and GP II, were obtained. Their chemical composition was elucidated from GC- and amino acid analyses as well as proton- and ^{13}C-nuclear magnetic resonance; the structures are presented in Fig. 1.

The glycan chain of GP I contains trisaccharide repeating units of α- and β-linked rhamnosyl residues and the chain consists of ca. 50 repeats (Christian et al. 1986). GP II is composed of tetrasaccharide repeating units with an apparent molecular weight of ca. 14,000. It consists of two di-N-acetylmannuronic acids, one N-acetylglucosamin and one glucose and bears negative charges from the uronic acids (Messner et al. 1987).

The glycan chain of *Clostridium thermohydrosulfuricum* L111-69 is composed of disaccharide repeatings of α-mannosyl- and α-rhamnosyl

Table 1. Carbohydrates on S-Layers of Eubacteria

Organism	Components
Gram-negative eubacteria	
Treponema sp., E-21 Masuda and Kawata (1982)	Neutral sugars, amino sugars (11%); PAS[a] (-)
Treponema phagedenis Masuda and Kawata (1986)	Neutral sugars, amino sugars (11%); PAS (-)
Campylobacter fetus, Winter et al. 1978	23D Hexose, pentose, methylpentose (3.8%)
Myxococcus xanthus, DK1050 Maeba (1986)	Neutral sugars, hexosamines, uronic acids (~13.5%); PAS (+)
Gram-positive eubacteria	
Bacillus brevis, S1 Abe et al. (1983)	PAS (+)
Bacillus sphaericus Word et al. (1983)	PAS (+)
Bacillus aneurinolyticus Abe and Kimoto (1984)	PAS (+)
Bacillus stearothermophilus, NRS 2004/3a Christian et al. (1986) Messner et al. (1987)	Neutral sugars, hexosamins, uronic acids (~5%) PAS(+)(see text)
Clostridium symbiosum, HB25 König et al. (1985)	Sugars (8.3%); PAS (+)
Clostridium tyrobutyricum Hayes et al. (1984) Bergère et al. (1986)	Neutral sugars (3-4%); PAS (-)
Clostridium thermohydrosulfuricum L111-69 Sleytr and Thorne (1976) Christian et al. (1988)	Neutral sugars, amino sugars (~8%) PAS (+) (for details see text)
Clostridium thermosaccharolyticum D120-70 Sleytr and Thorne (1976)	Neutral sugars (~3%) PAS(+)
Desulfotomaculum nigrificans Sleytr et al. (1986b)	Neutral sugars (1.5-8%) PAS(+)

[a]Periodic acid-Schiff staining reaction on SDS-polyacrylamide gels

residues with an apparent molecular weight of ca. 24,000 (Christian et al. 1988).

After the structural description of the glycan chains we were particularly interested in the elucidation of the protein-carbohydrate linkage region. Information on the linkage type in a glycopeptide can be obtained from β-elimination and hydrazinolysis experiments. With GP I of

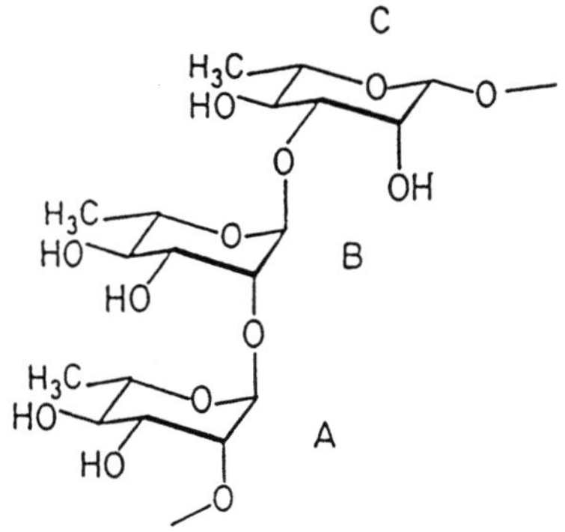

→2)-α-L-Rhap-(1→2)-α-L-Rhap-(1→3)-β-L-Rhap-(1→

Bacillus stearothermophilus NRS 2004/3a, GP I

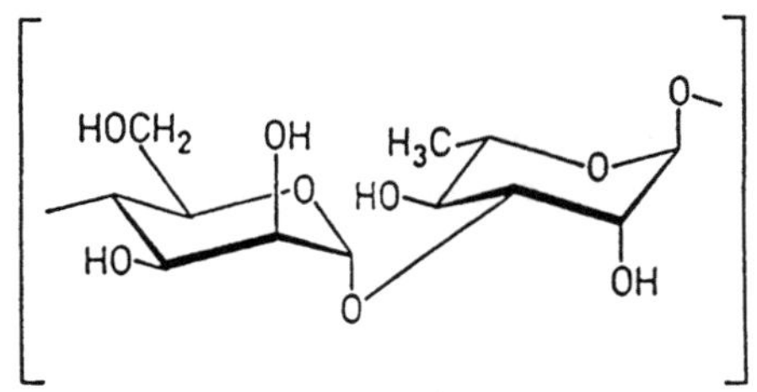

→4)-α-D-Manp-(1→3)-α-L-Rhap-(1→

Clostridium thermohydrosulfuricum L111-69

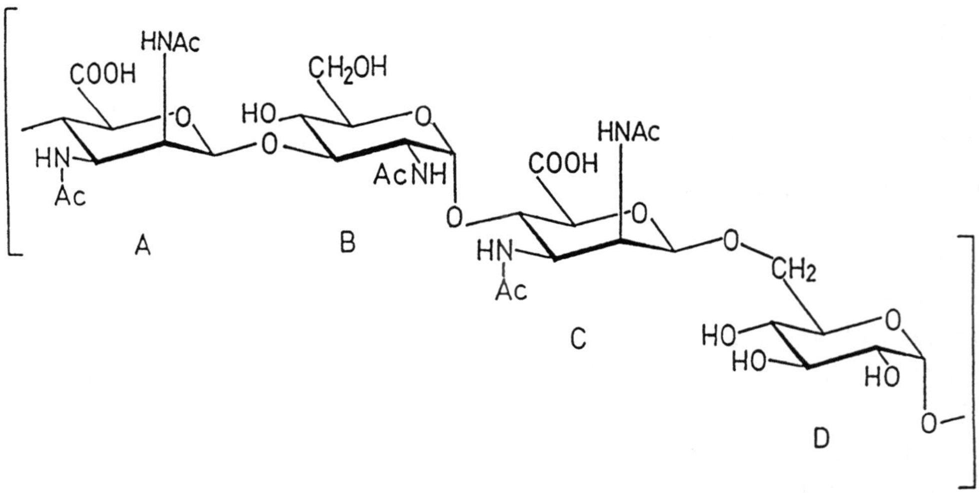

→4)-β-ManpA-2,3-diNAc-(1→3)-α-GlcpNAc-(1→4)-β-ManpA-2,3-diNAc-(1→6)-α-Glcp-(1→

Bacillus stearothermophilus NRS 2004/3a, GP II

Fig.1. Glycan structures of S-layer glycoproteins

Bacillus stearothermophilus NRS 2004/3a we have shown that cleavage of the molecule was possible only with hydrazine (Küpcü et al. 1984). This provides strong indication for an N-glycosidic linkage of the glycan to the peptide in GP I. Recently it could be shown that the polyrhamnan is N-glycosidically linked to asparagine in this molecule. This type of linkage has not yet been observed in archaebacterial or eukaryotic glycoproteins (Messner and Sleytr 1988).

From our data we conclude that glycoproteins exist on eubacteria and are not restricted to archaebacterial species (Mescher 1981, Kandler 1982). These findings led us to possible speculations about the biological functions of the carbohydrate portion in S-layers. In contrast to prokaryotes considerable knowledge has accumulated on eukaryotic glycoproteins. Potential functions could be altered cell-cell and cell-matrix recognition, protection against proteolytic degradation, changing of the physicochemical properties of the protein, or improvement of the natural survival under competitive conditions (Sharon 1984, Olden et al. 1985, West 1986, Sleytr et al. 1986b). These functions could also be relevant for prokaryotic cells. Protein glycosylation is an energy-expensive process to the organism because the carbohydrate moieties are secondary gene products necessitating the synthesis of specific enzymes. Therefore the ability for glycosylation should have provided advantages in selection during the evolution.

References

Abe M, Kimoto M (1984) Distribution of two types of regular array particles in the cell wall of *Bacillus aneurinolyticus* (Kimura et Aoyama). Microbiol Immunol 28:841-846

Abe M, Kimoto M, Yoshii Z (1983) Structural and chemical characterization of macro molecular arrays in the cell wall of *Bacillus brevis* S1. FEMS Microbiol Lett 18:263-267

Bergère J-L, Hayes H, Commissaire J (1986) Major protein components in the cell envelope of *Clostridium tyrobutyricum.* Ann Inst Pasteur Microbiol 137B:271-282

Christian R, Schulz G, Unger FM, Messner P, Küpcü Z, Sleytr UB (1986) Structure of a rhamnan from the surface layer glycoprotein of *Bacillus stearothermophilus* strain NRS 2004/3a. Carbohydr Res 150:265-272

Christian R, Messner P, Weiner C, Sleytr UB, Schulz G (1988) Structure of a glycan from the surface-layer glycoprotein of *Clostridium thermohydrosulfuricum* strain L111-69. Carbohydr Res (in press)

Hayes H, Commissaire J, Bergère J-L (1984) A high-molecular-mass cell wall protein released from *Clostridium tyrobutyricum* by heat treatment. FEBS Lett 174:284-288

Kandler O (1982) Cell wall structures and their phylogenetic implications. Zentralbl Bakteriol Hyg I Abt Orig C3:149-160

Kandler O, König H (1985) Cell envelopes of archaebacteria In: Woese CR, Wolfe RS (eds) The bacteria, vol VIII. Academic Press London New York, pp 413-457

König H, Stetter KO (1986) Studies on archaebacterial S-layers. Syst Appl Microbiol 7:300-309
König H, Buckel W, Langworthy TA (1985) Ultrastructure of the cell envelope and amino acid composition of the murein of *Clostridium symbiosum.* FEMS Microbiol Lett 30:283-288
Küpcü Z, März L, Messner P, Sleytr UB (1984) Evidence for the glycoprotein nature of the crystalline cell wall surface layer of *Bacillus stearothermophilus* strain NRS 2004/3a. FEBS Lett 173:185-190
Lechner J, Wieland F, Sumper M (1985) Transient methylation of dolichyl oligosaccharides is an obligatory step in halobacterial sulfated glycoprotein biosynthesis. J Biol Chem 260:8984-8989
Maeba PY (1986) Isolation of a surface glycoprotein from *Myxococcus xanthus.* J Bacteriol 166:644-650
Masuda K, Kawata T (1982) Isolation, properties, and reassembly of outer sheath carrying a polygonal array from an oral treponeme. J Bacteriol 150:1405-1413
Masuda K, Kawata T (1986) Isolation and structural and chemical characterization of outer sheath carrying a polygonal array from *Treponema phagedenis* biotyp Reiter. Microbiol Immunol 30:401-411
Mescher MF (1981) Glycoproteins as cell-surface structural components. TIBS 6:97-99
Mescher MF, Strominger JL (1976) Purification and characterization of a prokaryotic glycoprotein from the cell envelope of *Halobacterium salinarium.* J Biol Chem 251:2005-2014
Messner P, Sleytr UB (1988) Asparaginyl-rhamnose: A novel type of protein-carbohydrate linkage in a eubacterial surface-layer glycoprotein. FEBS Lett (in press)
Messner P, Sleytr UB, Christian R, Schulz G, Unger FM (1987) Isolation and structure of a diacetamidodideoxyuronic acid-containing glycan chain from the S-layer glycoprotein of *Bacillus stearothermophilus* NRS 2004/3a. Carbohydr Res 168:211-218
Olden K, Bernard BA, Humphries MJ, Yeo T-K, White SL, Newton JA, Bauer HC (1985) Function of glycoprotein glycans. TIBS 10:78-82
Paul G, Lottspeich F, Wieland F (1986) Asparingyl-N-acetyl galactosamine. Linkage unit of halobacterial glycosaminoglycan. J Biol Chem 261:1020-1024
Sharon N (1984) Glycoproteins. TIBS 9:198-202
Sleytr UB, Messner P (1983) Crystalline surface layers on bacteria. Annu Rev Microbiol 37:311-339
Sleytr UB, Thorne KJI (1976) Chemical characterization of the regular surface layers of *Clostridium thermosaccharolyticum* and *Clostridium thermohydrosulfuricum.* J Bacteriol 126:377-383
Sleytr UB, Messner P, Sára M, Pum D (1986a) Crystalline envelope layers in Archaebacteria. Syst Appl Microbiol 7:310-313
Sleytr UB, Sára M, Küpcü Z, Messner P (1986b) Structural and chemical characterization of S-layers of selected strains of *Bacillus stearothermophilus* and *Desulfotomaculum nigrificans.* Arch Microbiol 146:19-24
Sumper M (1987) Halobacterial glycoprotein biosynthesis. Biochim Biophys Acta 906:69-79
West CM (1986) Current ideas on the significance of protein glycosylation. Mol Cell Biochem 72:3-20
Wieland F (1988) The cell surfaces of halobacteria. CRC Crit Rev Microbiol (in press)
Wieland F, Heitzer R, Schaefer W (1983) Asparaginylglucose: A novel type of carbohydrate linkage. Proc Natl Acad Sci USA 80:5470-5474
Wieland F, Lechner J, Sumper M (1986) Iduronic acid: constituent of sulphated dolichyl phosphate oligosaccharides in halobacteria. FEBS Lett 195:77-81
Winter AJ, McCoy EC, Fullmer CS, Burda K, Bier PJ (1978) Microcapsule of *Campylobacter fetus:* Chemical and physical characterization. Infect Immun 22:963-971
Word NS, Yousten AA, Howard L (1983) Regularly structured and non-regularly structured surface layers of *Bacillus sphaericus.* FEMS Microbiol Lett 17:277-282

The Surface Layers of *Lampropedia hyalina*

J.W. Austin and R.G.E. Murray
Department of Microbiology and Immunology
The University of Western Ontario
London, Ontario
CANADA N6A 5C1

Lampropedia hyalina is a Gram-negative, aerobic, chemoorganotrophic coccus (Seeley 1974) that is distinguished by growth as a pellicle formed of vast flat sheets of cells arranged in fourfold symmetry (Pringsheim 1955, Starr and Skerman 1965) (Fig.1). These sheets arise by synchronous division of the cells, with each division alternating at right angles to the previous division to form square arrays (Murray 1963, Kuhn and Starr 1965). Cells within the sheet are held together by slender fibres, which are disposed between adjacent cells, as well as between the outer membrane and RS-layers (Chapman et al. 1963).

The sheets of cells in a square array appear to be cut into tablets of 32, 64, 128 or more cells. This is a result of the RS-layers of *Lampropedia hyalina* enclosing entire tablets of cells (Chapman et al. 1963), thereby subdividing the sheet. The outer RS-layer, or punctate layer, consists of flask-shaped units connected by Y-linkers and topped with a projecting spine (Murray 1963). The punctate layer is assembled onto an inner perforate layer (Fig.2), which is connected to the outer membrane by the fibres of the intercalated layer (Murray 1963, Chapman et al. 1963, Pangborn and Starr 1966). We have isolated and characterized the perforate layer (Austin and Murray submitted). Studies of the more complex, and delicate, punctate layer are ongoing and the results presented here are preliminary.

The Perforate Layer

The perforate layer was isolated from cell envelopes as large sheets by solution of the adherent membranes and punctate layer in 2% sodium dodecyl sulphate, followed by digestion of the peptidoglycan with lysozyme (200 µg/ml). The predominant feature of the layer was the p6 symmetry of the stain-filled holes in the sheet (Fig.3a). The centre-to-centre repeat of the arrangement was 13.5 nm, and sections through the layer showed the structure to be approximately 4 nm thick.

SDS polyacrylamide gel electrophoresis of the soluble perforate layer showed a single band, migrating at a position corresponding to 31.5 kDa (Fig.3b). Amino acid analysis of the component polypeptide revealed a high proportion of glutamic and aspartic acid. When the "ratio 3" was calculated, a value of 1.16 was obtained, typical of an extrinsic protein.

The Punctate Layer

Thin sections of tablets and negatively stained preparations of the RS-layers showed that the perforate layer and the punctate layer were closely associated, but did not identify precisely where the junctions were sited. The punctate layer consisted of 40-nm-high spines with apparently hollow centres, joined at the expanded middle of each spine, giving lattice constant of 23 nm. Negative stains of sheets of the punctate layer stripped from the cells by shearing in a blender demonstrated the hexagonal arrangement of the subunits joined by Y-shaped linking elements (Fig.4).

The punctate layer was extracted from cell envelopes by guanidine HCl and urea. SDS polyacrylamide gel electrophoresis of the soluble protein indicated a high molecular weight band at 240 kDa and additional bands migrating in the range between 50 kDa and 70 kDa (Fig.5). The perforate layer was solubilized at the highest concentrations of guanidine HCl and urea used.

Incorporation of alcian blue into the fixative for thin sections increased the staining of the punctate layer, and demonstrated amorphous material extending 20 to 40 nm outwards from the punctate layer (Fig.6). This material is presumed to be acidic exopolysaccharide. When used to stain SDS polyacrylamide gels, alcian blue bound to the 240 kDa polypeptide released from cell envelopes by guanidine HCl and urea. This suggests that the 240 kDa polypeptide has a polyanionic nature, and may be a glycoprotein or sulphated glycoprotein.

Incubation of the perforate and punctate layers in 3M guanidine HCl for short periods of time (15 min) at room temperature selectively solubilized the punctate layer, leaving the perforate layer intact. When this mixture of soluble punctate protein and intact perforate layer was dialyzed versus 10 nM $CaCl_2$ or 10 mM $SrCl_2$ the punctate layer reassembled onto the perforate layer. Dialysis against distilled water resulted in very limited reassembly, in the form of small patches. The

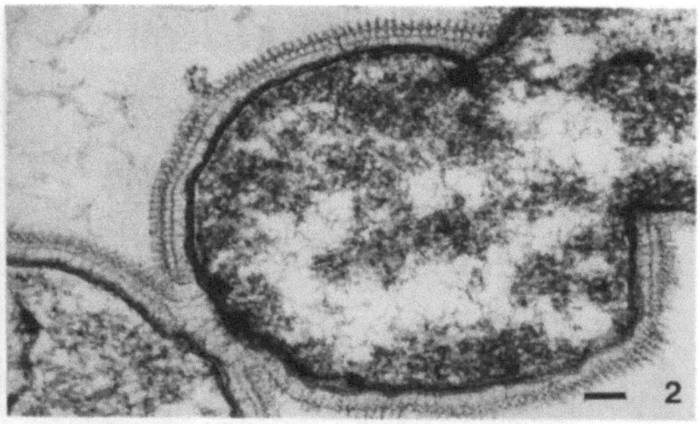

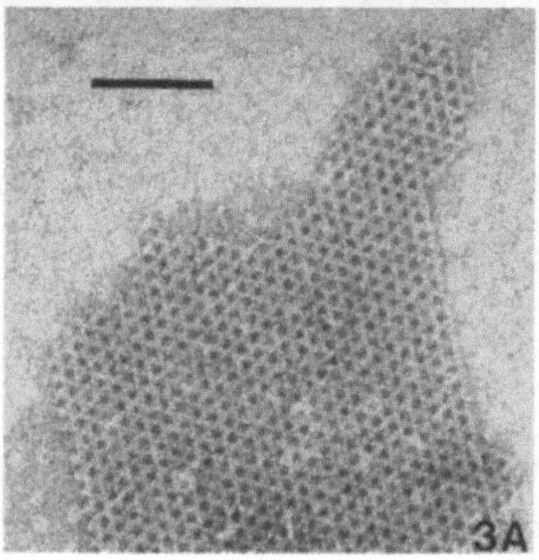

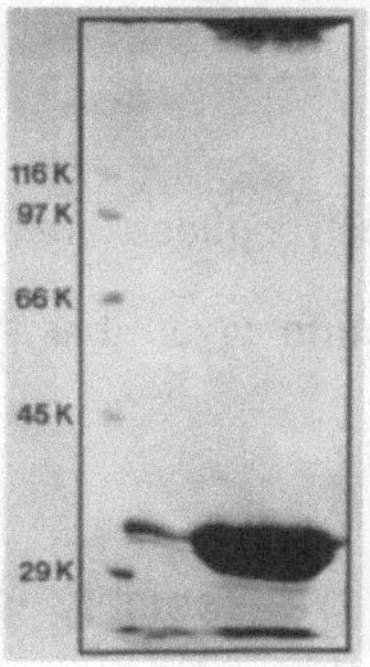

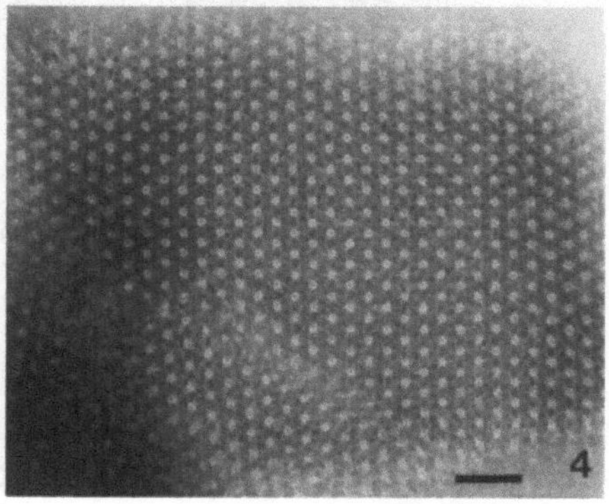

Fig.1. Phase contrast micrograph of *Lampropedia hyalina* (x 400)
Fig.2. Thin section showing the layers of the cell envelope (bar = 100 nm)
Fig.3. (a) Perforate layer negatively stained with 1% phosphotungstate (bar = 100 nm); (b) SDS polyacrylamide gel of the soluble perforate layer
Fig.4. Punctate layer stained with 1% ammonium molybdate and 0.1% glycerol (bar = 100 nm)

reassemblies formed by dialysis in distilled water were made soluble by subsequent dialysis against 50mM tetrasodium EDTA in 20 mM Tris pH 7.5, suggesting that calcium is required for reassembly of the punctate layer.

The RS-layer of *Flexibacter polymorphus* is reported to consist of four different polypeptides (Ridgeway and Lewin 1983). This is the only reported case of a single RS-layer with units consisting of more than one type of polypeptide. The RS-layers of both *Flexibacter polymorphus* and *Lampropedia hyalina* are complex and delicate structures. We hope to determine if the punctate layer consists of more

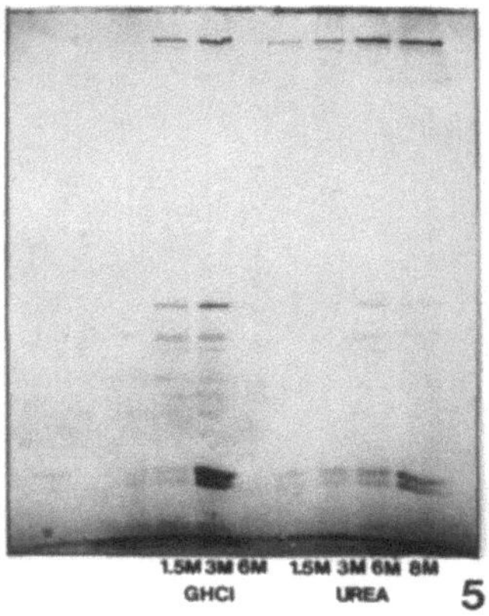

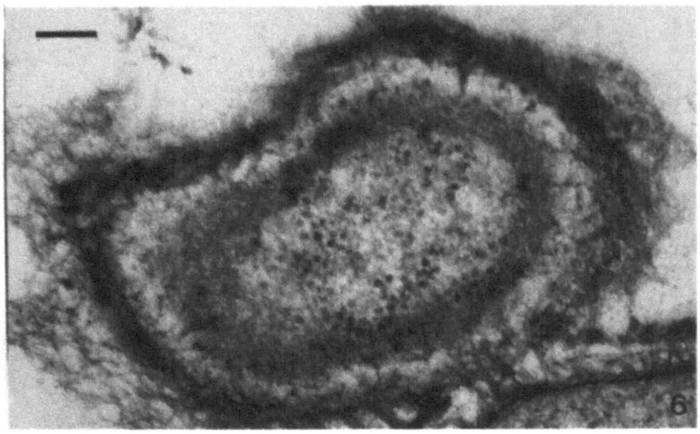

Fig.5. SDS polyacrylamide gel of polypeptides released from cell envelopes by guanidine HCl and urea
Fig.6. Binding of alcian blue to the punctate layer (bar = 100 nm)

than one type of polypeptide by reassembly studies, and the location of each of the component polypeptides may be determined by immunoelectron microscopy using antibodies against each of the components.

Acknowledgement

This research was supported by the Medical Research Council of Canada.

References

Chapman JA, Murray RGE, Salton MRJ (1963) The surface anatomy of *Lampropedia hyalina.* Proc R Soc London Ser B 158:498-513
Kuhn DA, Starr MP (1965) Clonal morphogenesis of *Lampropedia hyalina.* Arch Microbiol 52:360-375
Murray RGE (1963) Role of superficial structures in the characteristic morphology of *Lampropedia hyalina.* Can J Microbiol 9:593-600
Pangborn J, Starr MP (1966) Ultrastructure of *Lampropedia hyalina.* J Bacteriol 91:2025-2030
Pringsheim EG (1955) *Lampropedia hyalina* Schroeter 1886 and *Vannielia aggregata* n.g., n.sp., with remarks on natural and organized colonies in bacteria. J Gen Microbiol 13:285-291
Ridgeway HF, Lewin RA (1983) Subunit composition of goblet-shaped particles from the cell wall of *Flexibacter polymorphus.* Can J Microbiol 29:1689-1693
Seeley HW Jr (1974) Genus *Lampropedia* Schroeter 1886, In: Buchanan RE, Gibbons NE (eds), Bergey's manual of determinative bacteriology, 8th edn. Williams and Wilkins, Baltimore, pp 440-441
Starr MP, Skerman VBD (1965) Bacterial diversity: The natural history of selected morphologically unusual bacteria. Annu Rev Microbiol 19:407-454

A Glycoprotein Surface Layer Covers the Pseudomurein Sacculus of the Extreme Thermophile *Methanothermus fervidus*

E. Nußer[1], E. Hartmann[1], H. Allmeier[1], H. König[1], G. Paul[2] and K.O. Stetter[1]

[1]Lehrstuhl für Mikrobiologie and [2]Lehrstuhl für Biochemie I
Universität Regensburg, Universitätsstr. 31
D-8400 Regensburg
FRG

With a temperature maximum of 97°C, *Methanothermus* is the most thermophilic methanogenic genus. Only members of this archaebacterial genus possess a double-layered cell envelope composed of a pseudomurein sacculus and an S-layer (Fig.1a; Stetter et al. 1981). Each layer could be specifically removed: the S-layer with detergent or pronase (Stetter et al. 1981) and the pseudomurein sacculi of freshly harvested cells with a pseudomurein endopeptidase from *Methanobacterium wolfei* (Fig.1d). The cells are bipolar polytrichous flagellated. Up to 10 flagella with a length of 8 - 10 µm and a thickness of 13 - 15 nm could be found at each pole. The channels at the two poles of the pseudomurein sacculi are probably the passage holes of the flagella, since the diameter corresponds with the thickness of the flagella (Fig.1a,b). The thickness and the rigidity of the cell envelopes varied with the culture conditions. The cell envelope layers became thicker and more rigid in autotrophically than in heterotrophically grown cells. The thickness of the pseudomurein and the S-layer increased from 14 nm and 10 nm to 22 nm and 16 nm, respectively.

The S-layer glycoprotein could be solubilized by extracting whole cells with trichloroacetic acid (4%, 30 min, room temperature). The extracts were nearly free of protein impurities. Coextracted low molecular weight compounds were removed by reversed-phase chromatography (ProRPC HR 5/10; Pharmacia). Aqueous formic acid (10%; Heukeshoven and Dernick 1984) with subsequently increasing concentrations (0-85%) of methanol and isopropanol was used as eluant. The protein eluted at an isopropanol concentration of 85% (Fig.1e). When the eluted protein was subjected to SDS polyacrylamide gel electro-

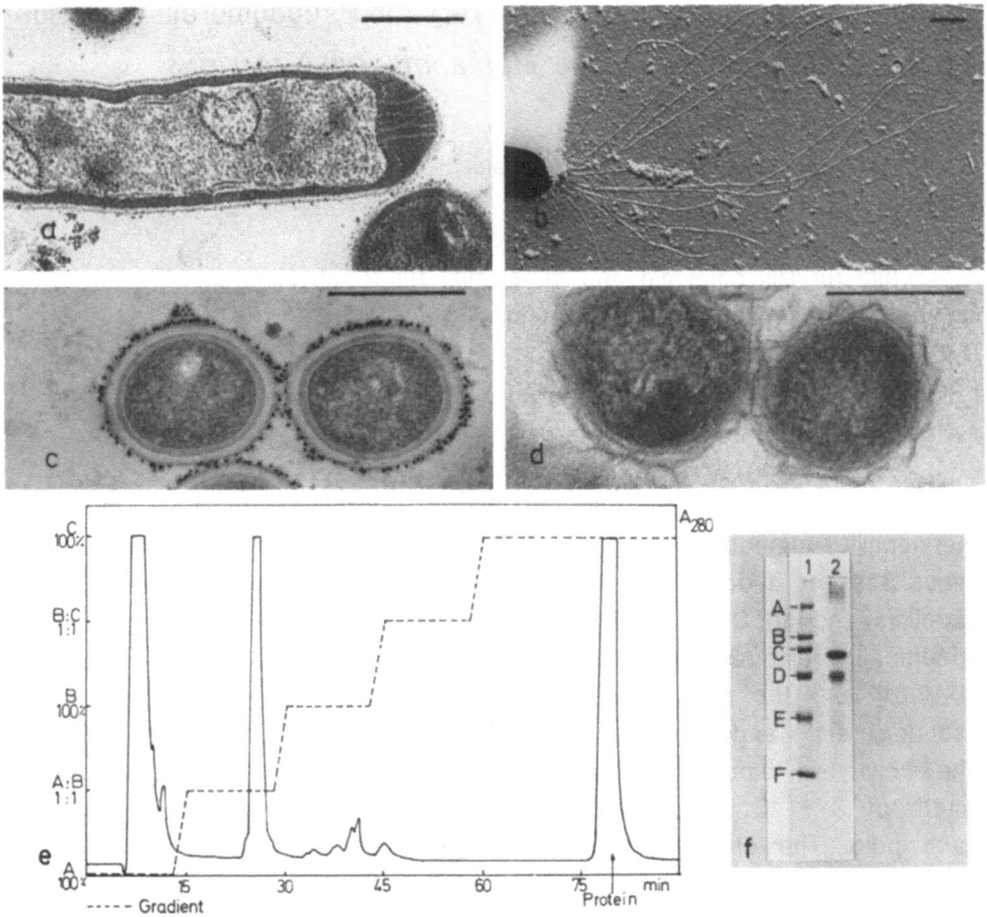

Fig.1.
Morphology of the cell envelope and isolation of the S-layer glycoprotein of *Methanothermus fervidus*
(a) - (d) Electron micrographs. Bar = 0.3 µm. (a) Thin longitudinal section. (b) Platinum-shadowed cell with flagella. (c) Thin cross-sections with immunogold labelling of the S-layer glycoprotein. (d) Thin cross-sections of cells treated with pseudomurein endopeptidase. (e) Elution profile of the trichloroacetic acid extract of whole cells separated on a reversed-phase column. Eluent A: 10% formic acid. Eluent B: 85% methanol in A (v/v). Eluent C: 85% isopropanol in A (v/v). (f) SDS-polyacrylamide gel of the trichloroacetic acid extract of cells of *Methanothermus fervidus*. 1. Molecular weight standards. (A = myosin, 205 kDa; B = β-galactosidase, 116 kDa; C = phosphorylase b, 97.5 kDa; D = bovine serum albumin, 66 kDa; E = ovalbumin, 45 kDa; F = carbonic anhydrase, 29 kDa), 2. cell extract

phoresis, two bands with molecular weights of 92,000 and 60,000 were obtained. Both bands were periodate-Schiff positive and reacted in immunoplots with antisera raised against the heavier band. After a preparative SDS gel electrophoresis followed by an additional reversed-phase chromatography each eluted band led to two bands on SDS polyacrylamide gels with the same molecular weights as above. This indicates that the two protein bands with different apparent molecular weights belong to one protein species. The localization of the isolated protein on the surface of the cells was proved by the immunogold-labelling technique (Fig.1c). The unusual behaviour against trichloroacetic acid is characteristic for the surface layer glycoproteins of *Methanothermus.* S-layer proteins of other extremely thermophilic species of the genera *Pyrodictium, Staphylothermus* and *Methanococcus* were not solubilized with trichloroacetic acid.

The chemical analysis revealed an unusual high amount of acidic amino acids (32 mol %; Table 1). Despite the comparable amount of acidic amino acids, the stability of the S-layer of *Methanothermus fervidus* must be controlled by other mechanisms than in *Halobacterium* (Wieland et al. 1982), because the S-layer is not disintegrated in distilled water in the presence of EDTA. The carbohydrate moiety accounts for about 17 mol %. It consists mainly of mannose and 3-O methylglucose in equal amounts, galactose and the N-acetylated amino sugars glucosamine and galactosamine. Pentoses, uronic acid or sulphated sugars could not be detected. 3-O-methylglucose was found for the first time as a constituent of an intact prokaryotic glycoprotein. It was identified as the corresponding alditol acetate derivative by its retention time on a Durabond capillary column and by combined gas-liquid and mass spectrometry after reduction of the C-1 carbon with $NaBD_4$. The isolation of glycopeptides after enzymatic hydrolysis of the S-layer protein has confirmed that the peptide is substituted by small oligosaccharides with a molecular weight of 2000.

During investigation of the biosynthesis of the carbohydrate chains sugar-1-phosphates and nucleotide activated derivatives of mannose, galactose, N-acetylglucosamine and N-acetylgalactosamine have been identified. The corresponding activated monomers of neither glucose nor 3-O-methylglucose could be isolated. They were found in nucleotide-activated oligosaccharides. Two kinds of uridine diphosphate-activated oligosaccharides were isolated; (a) oligosaccharides composed of galactose, mannose and 3-O-methylglucose, N-acetylglucosamine and N-acetylgalactosamine with N-acetylglucosamine at the reducing end and (b) oligosaccharides composed of mannose, galactose, glucose or 3-O-methylglucose with galactose at the reducing end. The occurrence of uridine diphosphate-activated oligosaccharides in the S-layer protein of *Methanothermus fervidus* is a remarkable difference to the biosynthesis of eukaryotic glycoproteins.

Table 1. Chemical composition of the S-layer glycoprotein (mol %)

Amino acids		Carbohydrates	
Asp	20.18	Man	7.80
Thr	5.93	Gal	0.71
Ser	5.43	3-O-Methylglc	6.98
Glu	6.97	GlcNAc	0.11
Pro	4.47	GalNAc	1.48
Gly	11.10	total:	17.08
Ala	2.74		
Cys	n.d.		
Val	3.53		
Ile	5.64		
Leu	2.95		
Tyr	1.18		
Phe	2.85		
His	1.48		
Lys	4.97		
Arg	2.97		
Met	0.52		
Trp	n.d.		
total:	82.91		

n.d. = not determined

References

Heukeshoven J, Dernick R (1984) Reversed-phase HPLC of virus proteins and other hydrophobic proteins. Chromatographia 19:95-100

Stetter KO, Thomm M, Winter J, Wildgruber G, Huber H, Zillig W, Janekovic D, König H, Wunderl S (1981) *Methanothermus fervidus,* a novel extremely thermophilic methanogen isolated from an Icelandic hot spring. Zentralbl Bakteriol Hyg I Abt Orig C2:166-178

Wieland F, Lechner J, Sumper M (1982) The cell wall glycoprotein of halobacteria: Structural, functional and biosynthetic aspects. Zentralbl Bakteriol Hyg I Abt Orig C3:161-170

The Structure, Chemistry and Physicochemistry of the *Methanospirillum hungatei* GP1 Sheath

T.J. Beveridge[1], M. Sára[2], D. Pum[2], G.D. Sprott[3], M. Stewart[4] and U.B. Sleytr[2]

[1]Dept.of Microbiology, C.B.S., University of Guelph, Guelph, Ontario N1G 2W1, CANADA
[2]Zentrum für Ultrastrukturforschung, Universität für Bodenkultur, A-1180 Wien, AUSTRIA
[3]Division of Biological Sciences, National Research Council of Canada, Ottawa, K1A OR6, CANADA
[4]MRC Molecular Biology Laboratory, Medical School, Cambridge CB2 2QH, UK

Methanospirillum hungatei is an archaebacterial methanogen which metabolizes $CO_2 + H_2$ into CH_4 under strictly anaerobic conditions. The bacterium exists as chains of rod-shaped cells which are packed into narrow filaments ca. nine cells long. A sheath is the filament's outermost structure which encloses the entire chain of cells; its rigidity gives the cells their shape and keeps them in chains (Beveridge et al. 1987).

The sheath is an extremely resilient proteinaceous structure, which resists all but the most potent chemical denaturants (Beveridge et al. 1985). This aids its isolation, since cells are extruded by 16.2 mM dithiothreitol, pH 10.0 treatment and lyzed by osmotic shock; the sheath remains and can be purified by multiple differential and sucrose isopycnic centrifugations (Sprott et al. 1983). We have recently discovered that a combination of 2% β-mercaptoethanol + pH=9.0 + 90°C for 30 min will dissociate the sheath into its constituent hoops (Fig.1) and liberate low MW "glue peptides" which cemented the hoops together (Sprott et al. 1986). Further treatment of the hoops in either 2% β-mercaptoethanol + 2% sodium dodecyl sulphate + pH = 11.1 + 90° C, or 0.05 M arginine-KOH buffer at pH=12.6 and 90°C, for 10-30 min will render them into three polypeptides (12, 24 and 45 kDa) which are resolvable by HPLC (Fig.2; Sprott et al. 1986).

The resilience of the sheath ensures that it is amenable to high resolution electron microscopy and negative stains reveal high order structure within the individual hoops which make up the intact structure (Fig.3). Small particles joined to the underlying fabric and spaced 2.8 nm apart cover its outer surface. Based on optical transforms, Shaw et al. (1985), suggested that these subunits had p 1 symmetry with a 2.8 nm unit cell (Fig.4). Stewart et al. (1985) proposed an alternative symmetry, based on faint 5.6 nm and major 2.8 nm reflections in optical and electron transforms and suggested that the symmetry was p 2 with a unit cell of 5.6 x 2.8 nm. Furthermore, high angle electron diffraction revealed a 0.47 nm repeat which indicated a cross-beta structure within the subunit (Stewart et al. 1985). Based on the dimensions of the unit cell, Sprott et al. (1986) suggested that two copies of the 24 kDa polypeptide form the 5.6 x 2.8 nm unit cell as a dimer.

So far, very little information is available about the surface physico-chemistry of the sheath. The proteins or glycoproteins of many S-layers, once assembled, have folded in a manner which makes them hydrophobic; the surfaces of these S-layers have few charge groups available (Sleytr and Messner 1983, Sára and Sleytr 1987, Messner et al. 1986). Titration of isolated sheath against increasing concentrations (2 μM to 2 mM) of uranyl acetate at pH=4.0 shows no detectable adsorption of the metallic ion; presumably at this low pH all available sites are protonated. Yet, at pH 6.8 (the pH at which the bacterium grows), polycationic ferritin (PCF, 10.4 mg per ml; Messner et al. 1986) adsorbed strongly (Fig.5). Stereo views of these preparations suggested that both inside and outside surfaces of the sheath were labelled and thin sections of glutaraldehyde/osmium tetroxide samples confirmed this. When the sheaths were chemically modified to neutralize carboxyl groups (carbodiimide activation followed by glycine methylester neutralization), the PCF adsorption was greatly reduced to background amounts which implicated these chemical sites as being responsible for the sheath's charge effect.

Since the sheaths are made up of hoops and since the hoops are of variable width along the sheath axis (Shaw et al. 1985, Beveridge unpublished), the packing order of the PCF particles depended on the hoop size. This was most apparent when the sheath was broken into fragments by sonication (3 x 60 sec sonications at maximum power; Branson Model B15) and PCF allowed to adsorb (Fig.6). Obviously, the

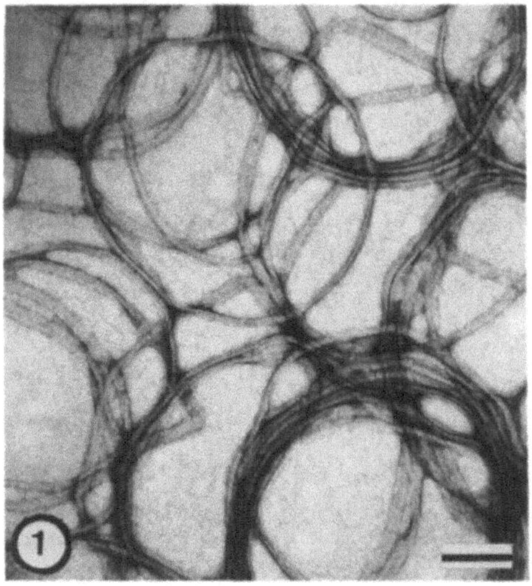

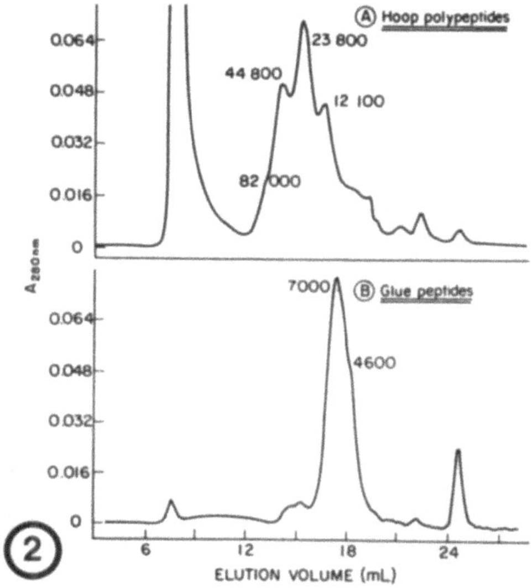

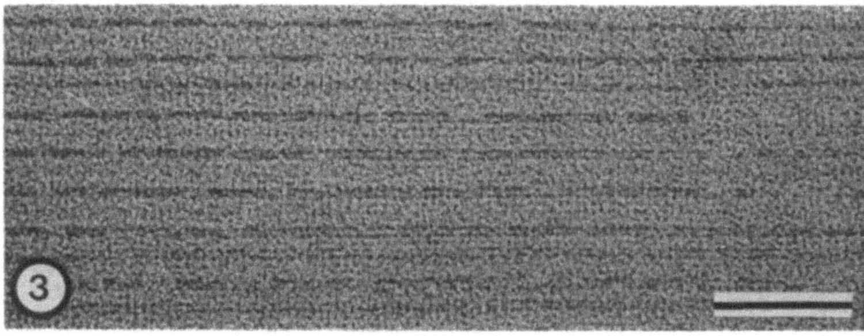

Fig.1. Negative stain (1% uranyl acetate) of hoops from the *Methanospirillum hungatei* sheath after β-mercaptoethanol/pH 10.0/90°C treatment for 30 min. Bar = 100 nm

Fig.2. HPLC profile of hoop proteins and low molecular weight peptides. Isolated sheath preparations lacking spacer plugs were treated twice with β-mercaptoethanol to separate hoops from low molecular weight peptides. (A) The hoops were solubilized in 0.05M L-arginine-KOH buffer at pH 12.6 (90°C for 10 min) and dialyzed for 1 h into the HPLC buffer. A sample of 1.7 mg protein was separated on a Superose 12 column, using 0.2M $NaHCO_3$, pH 7.8, as eluting buffer. The molecular weights shown in the figure represent averages of five separately solubilized preparations. (B) The column received 0.18 mg of low molecular weight peptides released by the first β-mercaptoethanol treatment. (Sprott et al. 1986)

Fig.3. High magnification of the intact sheath (1% uranyl acetate) which shows the high order structure within the constituent hoops. Bar = 50 nm

Fig.4. Optical transform of a sheath similar to that in Fig.3. The 2.8 nm reflections are clearly seen

Fig.5. Sheath treated by polycationic ferritin (PCF) to localize electronegative surface sites. Stained with 1% uranyl acetate. Bar = 200 nm

Fig.6. A single fragment of the sheath after sonication followed by PCF treatment. The packing order of the PCF particles and the major hoop striations are clearly seen. Stained with 1% uranyl acetate. Bar = 100 nm

Fig.7. Hoops after β-mercaptoethanol/pH 10.0/90°C treatment which breaks disulphide bonds and liberates small MW "glue peptides" (Sprott et al. 1986). PCF treatment completely coats the hoops on all sides and all edges. Stained with 1% uranyl acetate. Bar = 100 nm

Fig.8. An intact sheath after cytochrome *c* treatment to localize electronegative sites and stained with 1% uranyl acetate. A "fuzz" can be seen over the lattice. Bar = 100 nm.

Fig.9. Optical transform derived from Fig.8. The 2.8 nm reflections are apparent but now the 5.6 nm spacing is also very noticeable

PCF particle was too large a probe to discern accurately the location of the electronegative sites. It was also interesting to note that PCF did not adsorb to the broken edges of the sheath fragments (Fig.6); presumably these edges where not electronegative. Yet, when the sheath was chemically disassembled into hoops by the β-mercaptoethanol/ pH10/90°C method, PCF adsorbed to all edges and all sides of the hoops (Fig.7). It is probable that the breakage of disulphide bridges and the resulting dissolution of the "glue peptides" (Sprott et al. 1986) exposed previously unavailable electronegative sites.

Cytochrome c is a small (12.4 kDa), ellipsoidal (3.7 x 2.5 x 2.5 nm; Takano and Dickersen 1981) protein which, because of its high isoelectric point (pI = 10.8), is electropositive at neutral pH. When this molecule was used to probe electronegative sites on the sheath, negative stains gave the impression of adsorption (Fig.8). Optical diffraction of these images consistently revealed strong 5.6 nm and 2.8 nm reflections (Fig.9), and was strong evidence that the cytochrome c, by its binding, had altered the staining pattern so that the 5.6 nm repeat was more readily resolved . Computer reconstruction of these cytochrome c images is now being attempted.

References

Beveridge TJ, Stewart M, Doyle RJ, Sprott GD (1985) Unusual stability of the *Methanospirillum hungatei* sheath. J Bacteriol 162:728-737

Beveridge TJ, Harris BJ, Sprott GD (1987) Septation and filament splitting in *Methanospirillum hungatei*. Can J Microbiol 33:725-732

Messner P, Pum D, Sára M, Stetter KO, Sleytr UB (1986) Ultrastructure of the cell envelope of the archaebacteria *Thermoproteus tenax* and *Thermoproteus neutrophilus*. J Bacteriol 166:1046-1054

Sára M, Sleytr UB (1987) Charge distribution on the S-layer of *Bacillus stearothermophilus* NRS 1536/3c and importance of charged groups for morphogenesis and function. J Bacteriol 169:2804-2809

Shaw PJ, Hills GJ, Henwood JA, Harris JE, Archer DB (1985) Three-dimensional architecture of the cell sheath and septa of *Methanospirillum hungatei*. J Bacteriol 161:150-157

Sleytr UB, Messner P (1983) Crystalline surface layers on bacteria. Annu Rev Microbiol 37:311-339

Sprott GD, Shaw KM, Jarrell KF (1983) Isolation and chemical composition of the cytoplasmic membrane of the archaebacterium *Methanospirillum hungatei*. J Biol Chem 258:4026-4031

Sprott GD, Beveridge TJ, Patel GB, Ferrante G (1986) Sheath disassembly in *Methanospirillum hungatei* GP1. Can J Microbiol 32:847-854

Stewart M, Beveridge TJ, Sprott GD (1985) Crystalline order to high resolution in the sheath of *Methanospirillum hungatei*: a cross-beta structure. J Mol Biol 183: 509-515

Takano T, Dickersen RE (1981) Conformation change of cytochrome c. J Mol Biol 153: 79-94

Ultrastructural and Biochemical Studies of the Cell Sheath of *Methanothrix soehngenii*

P. Debeire[1], P. Pellerin[1], H.C. Dubourguier[1], B. Fournet[2], G. Prensier[1] and G. Albagnac[1]

[1] Institut National de la Recherche Agronomique. BP 39
F-59655 Villeneuve d'Ascq Cedex
[2] U.S.T.L. Laboratoire de Chimie Biologique
F-59655 Villeneuve d'Ascq Cedex
FRANCE

Glycoproteins are present in the cell envelopes of some Eubacteria (Küpcü et al. 1984) and Archaebacteria (Mescher and Strominger 1976, Wieland et al. 1980). Among Archaebacteria, the methanogens exhibit various types of cell envelopes (Kandler and König 1985), some of them being protein surface-layers associated with carbohydrates. *Methanothrix soehngenii* OPF (Zehnder et al. 1980) and *M. concilii* GP6 (Patel 1984) present similar characteristics and have been recently proposed as a synonym (Touzel et al. 1988).

The ultrastructure and the cell division of *Methanothrix soehngenii* strain FE used in this work are similar to those of *Methanothrix* strain GP6 extensively studied by Beveridge et al. (1986a,b). The cell sheath is made of hoops which can be destroyed by a lytic virus (Roustan et al. 1986). End-plugs are continuous to the sheath and oriented all along the filament. They are apparently made of a spiral arrangement and are double layered. Filament breakage occurs when multiple end-plugs are formed (Fig.1). In thin section, the sheaths showed a double layered structure (Fig.2). Only the outer one was stained by Thierry's reagent.

Purified sheaths were prepared from fresh cells of strain FE grown on polymannan-free medium in a 200-l fermentor. After French press disruption, nucleic acids were hydrolyzed with DNAse I and RNAse A. The pellet was washed twice with a Tris-HCl buffer containing 0.2% of Triton X 100 and 50 mM DTT. Purified sheaths were then dialyzed against distilled water and lyophilized.

They represented 10% of the cell dry weight and contained 81% of amino acids and 7% of carbohydrates. Compared with the strain OPF (König and Stetter 1986) and GP6 (Patel et al. 1986), the amino acid composition was quite similar except for glutamic acid and alanine. As in the two other strains, the cysteine content was low (0.6 mol per 100 residues). In contrast with OPF and GP6, rhamnose, fucose, ribose and glucose (5 : 1 : 1 : 1) were the dominant monosaccharides. Mannose was present at very low level whereas in OPF and GP6, it was dominant. This could be explained not only by adhesion to the sheaths of polymannans present in culture medium as observed with FE (unpublished results) but also by differences in structure or in biochemical techniques.

The purified sheaths were insoluble in polysaccharide extractants (DMSO, DEG, DMF, phenol 85^{0}). Performic acid oxidation or sodium hydroxide (0.05 N, 45^{0}C) had no effect on sheaths, suggesting that neither disulphide nor O-glycosidic linkages are involved in the sheath stability. At the opposite, sodium hydroxide (1 M, 100^{0}C) and anhydrous hydrazinolysis (100^{0}C, 20 h) led respectively to 85 and 96% of solubilization, suggesting that N-glycosidic or amido linkages are of major importance in the structural stability of the sheath.

Three major oligosaccharides were recovered after complete hydrazinolysis (100^{0}C, 20 h) followed by dry evaporation and solubilization in diluted acetic acid (1%). They were purified by permeation (TSK-SW) and ion exchange (Dowex 1 X 2) chromatography. The fraction of low molecular weight (3.3 kDa) contained mainly rhamnose, fucose and ribose (5 : 0.8 : 0.4). The neutral fraction (7.5 kDa) containing mainly the same monosaccharides (5 : 1.3 : 0.5) was partially characterized by GC-MS after permethylation. It was demonstrated that rhamnose was linked to other residues through positions 1, 2 and (or) 3 whereas fucose and ribose were linked through positions 1 and 3. The acidic fraction (12 kDa) contained mainly rhamnose and ribose (5 : 3). GC-MS studies showed that the linear structure was made of rhamnose residues linked through positions 1 and 2 and of ribose linked through positions 1 and 3, these two monosaccharides being also present as non-reducing end groups. Branching points were observed on position 2 of ribose. Further characterization is under progress.

Short-time hydrazinolysis (5 min, 95^{0}C) followed by two serial permeation chromatography (AcA34 and Trisacryl GF 2000) led to a single homogeneous peak (>1000 kDa) containing carbohydrates and

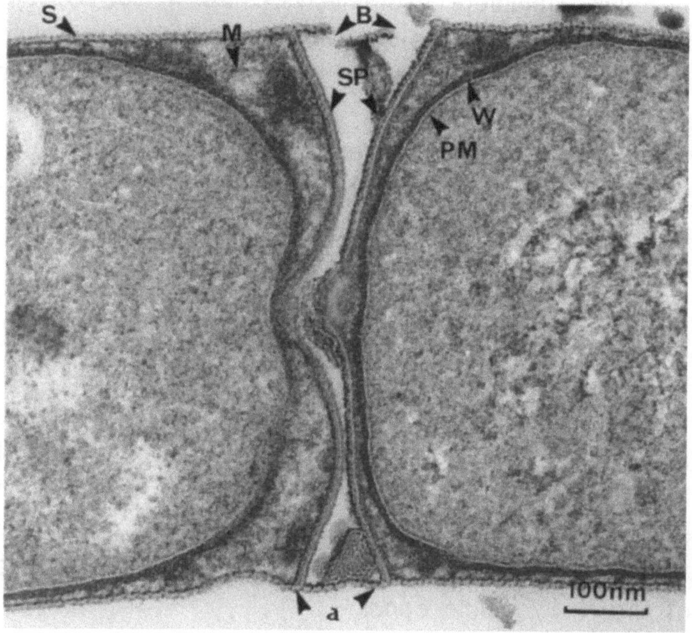

Fig.1. Thin section showing filament breakage (B) when two spacer-plugs (SP) are formed. Note the oriented continuity *(a)* of the sheath *(S)* with the spacer-plugs. The matrix *(M)* material is included between the sheath and a "cell wall" *(W)* veil surrounding the plasma membrane *(PM)*

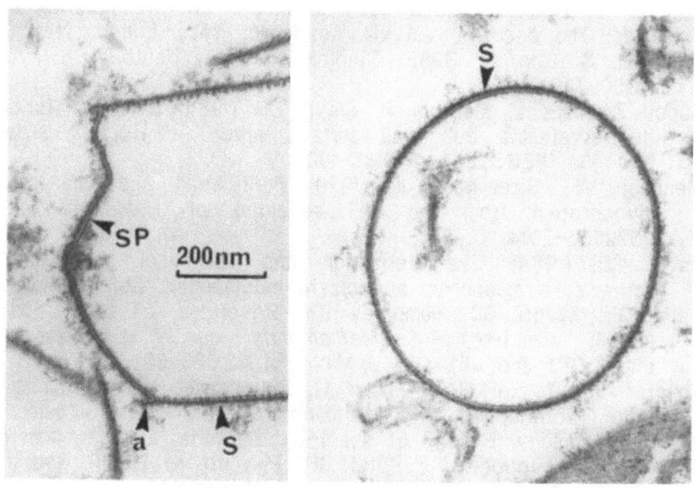

Fig.2. Thin section of purified sheaths. Note the double layered structure in *(S)*, the continuity *(a)* with the spacer-plug *(SP)*

proteins. Complete hydrazinolysis (20 h, 105°C) of this peak allowed the isolation of two major oligosaccharides of which elution patterns and composition corresponded to the acidic and neutral oligosaccharides previously described.

In contrast to the resistance of the purified sheaths to proteolytic enzymes, the single homogeneous peak was sensitive to proteinase K and a purified glycopeptide was obtained. Its content in aspartic and cysteic residues (respectively 22.0 and 5.3 mol per 100 residues) were higher than in entire sheaths. Rhamnose, fucose and ribose (5 : 1.15 : 1.3) were the dominant monosaccharides of the glycan moiety.

These results strongly suggest the presence of glycoproteins in the cell sheath of *Methanothrix soehngenii.* The unusual stability of the sheath may be mainly due to covalent linkages (N-glycosidic or amido) between subunits, contrarily to *Methanospirillum hungatei* in which the stability has been proved to be due to disulfide linkages (Sprott et al. 1986). Isolation of this major glycopeptide allows a study on the nature of the N-glycosidic bonds.

References

Beveridge TJ, Patel GB, Harris BJ, Sprott GD (1986a) The ultrastructure of *Methanothrix concilii,* a mesophilic aceticlastic methanogen. Can J Microbiol 32:703-710

Beveridge TJ, Harris BJ, Patel GB, Sprott GD (1986b) Cell division and filament splitting in *Methanothrix concilii.* Can J Microbiol 32:779-786

Kandler O, König H (1985) Cell envelopes of Archaebacteria. In: Woese CR, Wolfe RS (eds) The bacteria, vol VIII. Academic Press, London New York, pp 413-458

König H, Stetter KL (1986) Studies on archaebacterial S-layers. System Appl Microbiol 7:300-309

Küpcü Z, März L, Messner P, Sleytr UB (1984) Evidence for the glycoprotein nature of the crystalline cell wall surface layer of *Bacillus stearothermophilus* strain NRS 2004/3a. FEBS Lett 173:185-190

Mescher MF, Strominger JL (1976) Purification and characterization of a prokaryotic glycoprotein from the cell envelope of *Halobacterium salinarium.* J Biol Chem 251:2005-2014

Patel GB (1984) Characterization and nutritional properties of *Methanothrix concilii* sp. nov., a mesophilic aceticlaste methanogen. Can J Microbiol 30:1386-1396

Patel GB, Sprott GD, Humphrey RW, Beveridge TJ (1986) Comparative analyses of the sheath structures of *Methanothrix concilii* GP6 and *Methanospirillum hungatei* strain GP1 and JF1. Can J Microbiol 32:623-631

Roustan JL, Touzel JP, Prensier G, Dubourguier HC, Albagnac G (1986) Evidence for a lytic bacteriophage for *Methanothrix* sp. In: Dubourguier et al. (eds) Progress in biotechnology. Biology of anaerobic bacteria. Elsevier, Amsterdam, pp 200-205

Sprott GD, Beveridge TJ, Patel GB, Ferrante G (1986) Sheath disassembly in *Methanospirillum hungatei* strain GP1. Can J Microbiol 32:847-854

Touzel JP, Prensier G, Roustan JL, Thomas I, Dubourguier HC, Albagnac G (1988) Description of a new strain of *Methanothrix soehngenii* and rejection of *M. concilii* as a synonym of *M. soehngenii.* Int J Syst Bacteriol 38:30-36

Wieland F, Dompert W, Bernhardt G, Sumper M (1980) Halobacterial glycoprotein saccharides contain covalently linked sulphate. FEBS Lett 120:110-114
Zehnder AJB, Huser BA, Brock TD, Wuhrmann K (1980) Characterization of an acetate-decarboxilating non-hydrogen-oxidizing methane bacterium. Arch Microbiol 124:1-11

The S-Layers of *Aquaspirillum serpens*

S.F. Koval, M.L. Kist and R.G.E. Murray

Department of Microbiology and Immunology
University of Western Ontario
London, Ontario
CANADA N6A 5C1

Aquaspirillum serpens is a Gram-negative, helical organism which is widely distributed in freshwater environments. Of the 14 strains of *A. serpens* in our culture collection, nine possess S-layers, all of which have hexagonal (p6) symmetry. The S-layers vary in centre-to-centre spacing of subunits and type of connectivity. The S-layer proteins of *A. serpens* strains MW5 and VHA have proven to be most suitable for structural and biochemical analyses. Comparative studies on these S-layers have provided useful information on the assembly of single (strain VHA) and double (strain MW5) S-layers. Some of the biochemical data referred to as "in preparation" is in a manuscript by Koval, Kist, Boivin and Murray.

Ultrastructure

The S-layer of strain VHA is an M_6C_3 type (Saxton and Baumeister 1986): i.e. the heavier domains of the single molecular species are arranged in hexagonal symmetry with the lighter domains providing "Y" linkers (Murray 1963). The unit cell has a centre-to-centre spacing of 14.5 nm (Fig.1). It is difficult to obtain sheets of the S-layer with well-preserved structure: often the "linkers" are distorted, and high resolution image processing is difficult. Dickson et al. (1986) were successful in preparing a few single-layered arrays of the S-layer protein free of the underlying outer membrane and obtained a three-dimensional reconstruction to a resolution of about 1.6 nm of negatively stained specimens. The structure consists of a core in the shape of a cup, with six projections arising from the rim of the cup to join adjacent subunits at the threefold symmetry axes.

The structure of the S-layers of strain MW5 has provided an intriguing problem. Negative staining of whole cells shows a predominant linear, or ribbed pattern, while a similar staining of cell envelopes often reveals patches of a hexagonal arrangement of subunits, joined by "delta" linkers (an M_6C_2 structure), with a centre-to-centre spacing of 16 nm (Fig.2) Thin sections of cells show that two S-layers are present. Stewart and Murray (1982) elegantly demonstrated that the linear pattern arose from the juxtaposition of two identical hexagonal layers, with one sheet staggered by one-half repeat along a hexagonal (1,0) lattice line relative to the other.

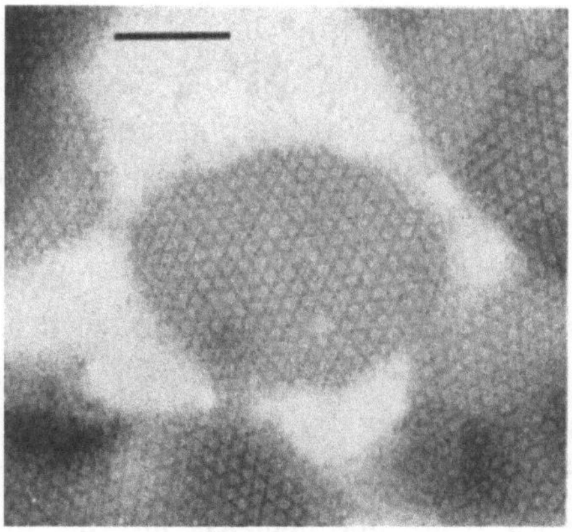

Fig.1. Negative stain (1% ammonium molybdate) of cell envelopes of *A. serpens* VHA. Bar = 100 nm

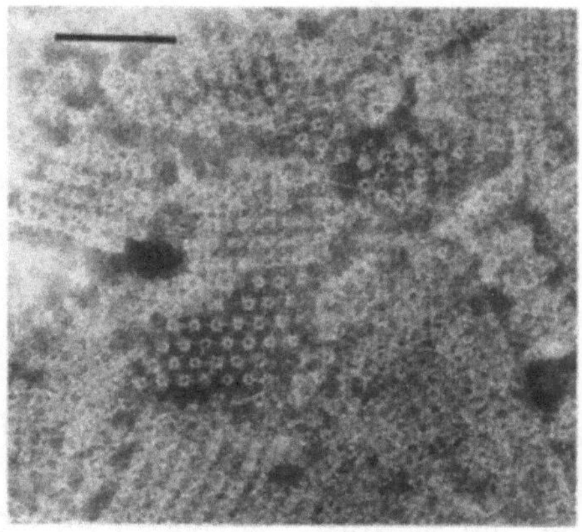

Fig.2. Negative stain (1% phosphotungstic acid) of S-layers of *A. serpens* MW5. Bar = 100 nm

Biochemical Characterization

The S-layer of strain VHA is composed of a single polypeptide (molecular weight 140,000), which does not contain carbohydrate. The S-layer protein has an isoelectric point of 4.45 (Koval et al. in prep.) and contains a high proportion of acidic and non-polar amino acids, with no cysteine and a low amount of methionine. Strain VHA requires calcium for the in vitro (Buckmire and Murray 1976) and in vivo (Koval and Murray 1985) assembly of the S-layer. This S-layer protein will not self-assemble. The S-layer protein isolated with 10 mM urea was shown (by a colorimetric assay) to contain tightly bound calcium (Koval et al. in prep.). Some degradation of the S-layer protein always occurs in guanidine or urea extracts. We are able to limit the amount of proteolytic degradation by the inclusion of 10 mM calcium during the isolation procedure (Koval et al. in prep.). Calcium is not required for synthesis of the S-layer protein (Koval and Murray 1985). In calcium-deficient medium an S-layer is not assembled on the cell surface and a small amount of protein is secreted into the medium. There must be a tight regulation of the synthesis, transport and assembly of the protein since large amounts are not secreted during growth, in the presence or absence of calcium. Preliminary studies indicate that the S-layer protein is not synthesized as a proprotein. Exponential phase cells were pretreated with CCCP prior to labelling of proteins with ^{35}S-methionine. The 140,000 polypeptide was immunoprecipitated with anti-140,000 IgG, but no higher molecular weight polypeptide was observed in the immunoprecipitates.

The outer and inner S-layers of strain MW5 are composed of two proteins with molecular weights of 125,000 and 150,000 respectively (Kist and Murray 1984). These proteins can be solubilized from cell envelopes either by extraction with 6 M urea (which solubilizes both proteins) or by a sequential extraction using buffer at pH 10.3 followed by a 6 M urea extraction. These S-layer proteins do not contain carbohydrate. Both proteins have an isoelectric point of 4.7 (Koval et al. in prep.). The amino acid profiles of the proteins are similar to each other and to that of the 140,000 protein of strain VHA. Strain MW5 also requires calcium for the in vitro reassembly of both S-layers (Kist and Murray 1984). The two proteins can either reassemble sequentially onto the cell surface, or self-assemble into planar sheets possessing the linear pattern. In addition, the 150,000 protein, by itself, can self-assemble to form hexagonal arrays. Biochemical and structural analyses have shown that this protein is isolated in a multimeric form, which resembles the circular units that comprise the intact hexagonal array (Koval et al. in prep.). The calcium-dependent step in the in vitro assembly of the 150,000 protein appears to be in the formation of hexagonal arrays from intact units. The S-layer proteins of strain MW5 also contained bound calcium (Koval et al. in prep.). However, attempts to minimize proteolytic degradation of these proteins during isolation by the presence of calcium at all stages of purification were not successful.

Relatedness

The biochemical relatedness of the S-layer proteins of strains VHA and MW5 was examined by two-dimensional peptide mapping (Koval et al. in prep.). The strain MW5 proteins shared approximately 12 peptides (out of a total of about 28) when digested with trypsin, and 8 peptides when digested with *Staphylococcus aureus* V8 enzyme. These proteins had four to five peptides in common with the strain VHA S-layer protein. Overall, the strain MW5 proteins appeared more similar to each other than did either to the strain VHA protein. Thus, although Stewart and Murray (1982) could not detect any structural differences (at a resolution of 2.5 nm) between the two S-layers of strain MW5, peptide mapping has shown that one S-layer protein is not a simple modification of the other.

The antigenic relatedness of the S-layer proteins was examined by the Western immunoblotting method using IgG isolated from antiserum raised against SDS-denatured proteins (Koval et al. in prep.). Anti-150,000 IgG (MW5) did not cross-react with the 125,000 (MW5) protein and reacted only with both the 150,000 (MW5) and 140,000 (VHA) proteins. Anti-140,000 IgG (VHA) bound to both the proteins of strain MW5, although greater cross-reactivity was shown with the 125,000 protein. This antigenic cross-reactivity would be expected, given the shared sequences demonstrated by peptide mapping, and also because the antibody was obtained by challenging rabbits with SDS-denatured protein.

The S-layer proteins of both strains are thus biochemically and antigenically related and it is intriguing to think of these as representing common structural and/or functional domains.

Acknowledgement

This research was supported by the Medical Research Council of Canada.

References

Buckmire FLA, Murray RGE (1976) Substructure and in vitro assembly of the outer, structured layer of *Spirillum serpens.* J Bacteriol 125:290-299

Dickson MR, Downing KH, Wu WH, Glaeser RM (1986) Three-dimensional structure of the surface layer protein of *Aquaspirillum serpens* VHA determined by electron crystallography. J Bacteriol 167:1025-1034

Kist ML, Murray RGE (1984) Components of the regular surface array of *Aquaspirillum serpens* MW5 and their assembly in vitro. J Bacteriol 157:599-606

Koval SF, Murray RGE (1985) Effect of calcium on the in vivo assembly of the surface protein of *Aquaspirillum serpens* VHA. Can J Microbiol 31:261-267

Murray RGE (1963) On the cell wall structure of *Spirillum serpens.* Can J Microbiol 9:381-392

Saxton WO, Baumeister W (1986) Principles of organization in S-layers. Mol Biol 187:251-253

Stewart M, Murray RGE (1982) Structure of the regular surface layer of *Aquaspirillum serpens* MW5. J Bacteriol 150:348-357

Image Analysis of the Two Patterns from Negatively Stained A-Layers of *Aeromonas salmonicida:* Correction for Lattice Distortions

M. Stewart[1] and T.J. Beveridge[2]

[1]Medical Research Council, Laboratory of Molecular Biology
Hills Rd., Cambridge CB2 2QH, ENGLAND
[2]Department of Microbiology, University of Guelph
Guelph, CANADA N1G 2W1

There is strong evidence linking the presence of an outer cell wall layer (the A-layer) to virulence of *Aeromonas salmonicida* strains (see Ishiguro et al. 1981). We have previously shown that negatively stained A-layer preparations show two different types of pattern, both of which have p4 symmetry in projection (Stewart et al. 1986), and raised the possibility that these patterns may reflect a structural transition in the A-layer that might alter its permeability. However, analysis of these patterns is not completely straightforward because the crystalline lattice is often not perfect. In one type of pattern this made Fourier-based methods difficult to apply and our previous work instead employed correlation averaging. Although this gave some enhancement of signal/noise it was difficult to assess how great this was and what the final resolution was. We have now used a more elaborate form of analysis, based on unbending the distorted lattices, to address these problems and thereby obtain a more reliable picture of the two A-layer patterns and their relationship to one another.

The methods we used have been described in detail (Henderson et al. 1986) and have been used previously on data from bacteriorhodpsin and acetylcholine receptor crystals. Broadly, these methods locate the position of unit cells by cross-correlation and then fit a bicubic spline to model the lattice distortion. Bilinear interpolation is then used to remove (or at least reduce) these distortions. Since a principal effect of disorder is to broaden spots, this processing sharpens spots, making them easier to observe above the background (and so detect if they are weak) and also increases their signal/noise ratio, thereby making it possible to obtain more accurate amplitude and phase information.

We monitored the progress of unbending by measuring the extent to which the peaks in the Fourier transform resembled those expected for a perfect lattice and also by the number of spots that had a signal/noise ratio greater than 2. Fig.1 shows the general effect on spot shape and strength for the two types of pattern and Table 1 compares numbers of spots, power and phase errors. These data show that correction of lattice distortions has significantly improved the quality of the Fourier data. In type I images, in which the level of lattice distortion did not appear very pronounced, there was still an increase in the number of spots detected. The overall power in these spots increased by about 20% and the amplitude-weighted phase residuals decreased by about half.

36	35	38	35	40	38	37	37	37
37	36	35	38	40	41	34	35	35
39	39	40	37	52	50	42	37	38
37	37	42	58	76	66	48	38	38
38	41	44	90	198	88	53	44	37
37	36	45	58	74	66	52	37	38
37	39	38	42	49	44	37	39	35
34	36	38	38	38	40	35	35	34
35	34	36	37	41	38	38	33	35

Type I patterns before unbending

39	38	39	39	38	37	39	38	37
39	37	39	40	40	42	38	39	36
42	38	41	46	50	48	43	36	40
40	39	47	54	73	58	41	41	37
42	45	53	100	163	84	51	45	38
34	38	53	58	87	63	47	42	38
37	40	44	45	51	42	36	41	32
39	38	40	40	44	40	40	36	35
41	39	39	42	35	36	39	34	40

Type II patterns before unbending

32	31	34	36	34	31	32	29	31
31	34	29	32	38	32	33	33	30
33	34	36	34	45	37	32	33	33
34	35	35	41	99	51	37	35	33
31	33	39	68	213	63	42	39	34
31	32	35	51	68	46	39	34	32
34	23	34	40	41	41	36	31	36
32	31	35	34	34	33	29	31	30
31	35	30	31	34	32	30	31	31

Type I patterns after unbending

39	37	38	38	37	35	36	33	30
39	36	40	36	37	35	32	30	31
39	39	37	38	44	34	36	33	34
38	38	38	46	62	49	35	34	35
39	45	45	99	194	71	42	36	34
39	40	42	52	73	41	37	35	33
38	36	40	43	45	40	32	33	35
39	41	41	36	39	37	33	33	36
37	39	38	34	37	35	36	30	32

Type II patterns after unbending

Fig.1 .Square root of summed squared amplitudes in a 9 x 9 pixel box centred on each lattice point in the Fourier transform for all spots from three images of each type of pattern. Unbending has increased the overall peak height and sharpness, particularly for type II patterns

The effect of correcting type II images was more dramatic. The raw images were so distorted that Fourier-based processing was not really feasible: most spots were weak and rather diffuse. However, after processing, the number of spots not only increased, but their sharpness increased dramatically, as did their power, while again the amplitude-weighted phase residual decreased markedly.

Table 1. Illustration of data on the two pattern types before and after correction

	Before correction		After correction	
	Type I	Type II	Type I	Type II
Number of spots	21	16	25	22
p4 Power (arbitrary units)	3580	2673	4320	3966
Phase residual (degrees)	10.6	10.9	5.6	5.4

Figure 2 shows reconstructed images of the two types of pattern. The features broadly resemble those we identified using correlation averaging (Stewart et al. 1986). Both patterns contain two types of subunit on a centric lattice so that annular units alternate with those constructed from four lobes in a square planar configuration. In the Fourier-processed images, the weight of each type of subunit in each pattern is more uniform than in the reconstructions we obtained previously using correlation averaging, which adds confidence to our interpretation that both types of pattern are constituted from the same sorts of subunit. Also, as we observed earlier, the major difference between patterns derives from rotation of the square planar subunit to produce apparently larger pores in type II patterns. Thus, our more extensive analysis confirmed our previous result and has increased our confidence in both the morphology of the layer and the relationship between the two patterns observed.

As we noted earlier (Stewart et al. 1986), this relationship between the patterns could be consistent with a structural transition in the A-layer that might change its permeability, possibly in a manner analogous to gap junctions in eukaryotic cells. Such a change in permeability might have important consequences for the pathogenesis of *A. salmonicida.*

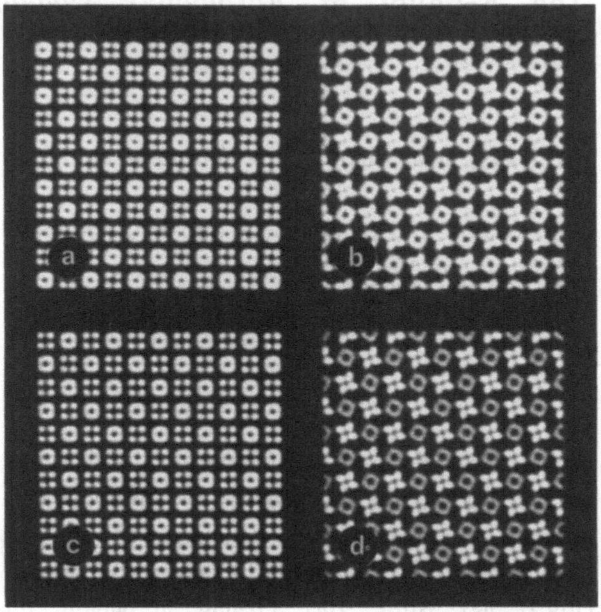

Fig.2. (a,b) Reconstructed images of the two types of pattern seen in negatively stained isolated A-layers after correction for lattice distortion using the method of Henderson et al. (1986). These reconstructed images closely resemble those (c,d) obtained by correlation averaging (Stewart et al. 1986) but the density of the two types of subunit is more similar between patterns. (a,c) type I pattern; (b,d) type II pattern

References

Henderson H, Baldwin JM, Downing KH, Lepault J, Zemlin F (1986) Structure of purple membrane from *Halobacterium halobium:* recording, measurement and evaluation of electron micrographs at 3.5A resolution. Ultramicroscopy 19:147-178

Ishiguro EE, Kay WW, Ainsworth T, Chamberlain JB, Austin RA, Buckley TJ, Trust TJ (1981) Loss of virulence during culture of *Aeromonas salmonicida* at high temperature. J Bacteriol 148:333-340

Stewart M, Beveridge TJ, Trust TJ (1986) Two patterns in the *Aeromonas salmonicida* A-layer may reflect a structural transformation that alters permeability. J Bacteriol 166:120-127

The A-Layer of *Aeromonas salmonicida:* Organization and Functions

W.W. Kay, B.M. Phipps, R.A. Garduño and T.J. Trust

Department of Biochemistry and Microbiology
University of Victoria
Victoria, British Columbia V8W 2Y2
CANADA

Since the discovery of bacterial S-layers in 1953, crystalline surface arrays composed of protein or glycoprotein subunits have been found in nearly all phylogenetic divisions of the eubacteria and archaebacteria, and in a few species of eukaryotic algae as well. Sleytr and Messner (1983) list over 390 strains in which the surface arrays have been identified (see also appendix). Thus the ubiquitous occurrence of S-layers amongst diverse genera and ecological niches suggests that S-layers fulfill important functions in bacteria. Pathogenic bacteria, such as *Aeromonas salmonicida,* present unique opportunities to investigate the function of the S-layer since such layers represent a specific interface between the organism and the environment - its host. *A. salmonicida* is a pathogen of a wide variety of fish and is the causative agent of the systemic disease furunculosis, a worldwide commercially important disease in cultured salmonids (Herman 1968). Virulent strains possess a cell surface protein array designated as the A-layer (Udey and Fryer 1978). Isogenic strains devoid of the A-layer protein (A-protein) are avirulent (Udey and Fryer 1978, Kay et al. 1981), thus establishing the central role of the layer as a virulence factor. It is the intent of this contribution to present the properties of the A-layer which contributes to its essential role in virulence.

Structure and Organization

From previous studies the A-layer of *A. salmonicida* can be seen to be composed of the 49,000 M_r A-protein, tetragonally arrayed, with the potential for conformational changes to this pattern. The following summarizes our current view of the organization of the A-layer relative to the cell surface. Our model depicts A-protein monomers comprised of two domains tetragonally arrayed according to the M_4C_4 model proposed by Baumeister and Engelhardt (1987). The A-layer completely covers the cell surface since anti-A-protein IgG, when fluorescein conjugated can be shown to completely surround the cell (Phipps et al. 1983). In addition, impermeant radiolabelling reagents, such as ^{125}I diazotized sulphanilic acid label only the A-protein (Kay et al. 1981). The A-protein alone comprises the subunit associations typically seen by electron microscopy. When purified free of LPS either from the surface of A^+ cells or as secreted A-protein it reconstitutes into the typical tetragonal array. The subunit associations are mediated by ionic linkages since treatment of A^+ cells with EGTA completely dissociates the tetragonal array but treatment with EDTA has little effect. Our current view is the A-protein associates with the core region of LPS and is anchored to the outer membrane of the cell via the LPS-lipid-A portion. Isogenic O-antigen-deficient LPS mutants of *A. salmonicida* secrete A-protein directly into the medium during growth (Ishiguro et al. 1984, Belland and Trust 1985). Association of A-protein with major outer membrane proteins could not be shown by various cross-linking studies. However, A-protein is within reach of a 26,000 M_r minor outer membrane protein as shown by cross-linking with dithiobis(succinimidyl)propionimidate.

Complement Resistance

The complement system of both homeothermic animals and poikilothermic teleost fish plays an important role in the protection against bacterial infection. In addition to the direct bactericidal reaction of complement it also has an opsonic function in the interaction between the bacteria and phagocytic cells (Frank 1979). However, since fish phagocytes appear to lack C3 receptors (Wrathmell and Parish 1980), the direct bactericidal activity of complement would seem to be most important.

A variety of *A. salmonicida* strains and isogenic mutants deficient in either A-protein, LPS or both were evaluated with respect to their resistance to the bactericidal activities of human, rabbit, and trout sera, immune trout sera, and immune rabbit sera which contained antibodies to both A-protein and LPS. Virulent strains of *A. salmonicida* displayed either higher or intermediate resistance to the bactericidal activity of complement both in the presence and absence of specific antibody. In normal sera, both A-protein and LPS were shown to contribute to serum resistance. In immune trout serum, the protection was conferred by A-protein. In fact, in immune trout serum containing high levels of antibody, all A^- strains were quickly killed, whereas there was no enhanced killing of their A^+ parents. Thus the A-layer must function as a protective physical barrier to the access of the lytic complement components to the cytoplasmic membrane.

Fish phagocytes appear to lack C3 receptors (Wrathmell and Parish 1980), which would apparently preclude opsonization by phagocytes. However, the chemistry of the A-layer surface may have been devised to overcome an opsonizing defect.

Cell Surface Hydrophobicity and Macrophage Association

The cell surface hydrophobicity of *A. salmonicida* is determined principally by the balance between the very hydrophobic A-protein and the very hydrophilic LPS components of the A-layer. When *A. salmonicida* (A^+) and A-layer-deficient (A^-) isogenic mutants were assayed for the ability to associate with both cultured rainbow trout macrophages and cultured mouse macrophages, the A-layer positive strains were invariably two to three times more active in their ability to associate with macrophages (Fig. 1).

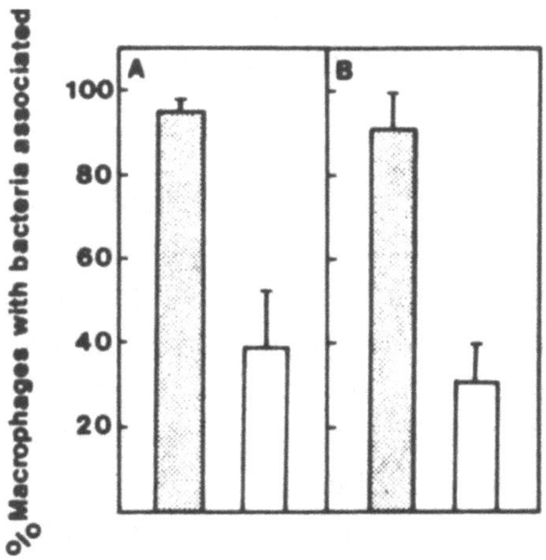

Fig.1. Association of A^+ *A. salmonicida* (shaded bars) and an A^- strain (open) bars) with (A) cultured rainbow trout macrophages, (B) cultured mouse macrophages.

Internalization of A^+ cells by macrophages was also indicated by the indirect fluorescence assay (IFAT) for these bacteria using unfixed preparations. Macrophage association by *A. salmonicida* appears to be specific since the ability of A^+ cells to bind to macrophages is not seen with other cell types.

At first it would seem suicidal for A^+ *A. salmonicida* cells to associate with macrophages; however, we have not been able to demonstrate macrophage killing of *A. salmonicida* after internalization of bacterial cells. This suggests that the macrophage becomes an unwilling accomplice in the rapid spread of the infection. At this time it is unknown whether the A-layer facilitates bacterial invasion of other tissues.

Immunoglobulin Binding

Another mechanism by which *A. salmonicida* may seek to evade the host immune defense system is through the binding of immunoglobulins. *A. salmonicida* cells containing the A-layer were found to specifically bind rabbit IgG and human IgM with relative high affinity (K_d's 1 x 10^{-6} M and 3.3 x 10^{-6} M respectively) (Fig.2).

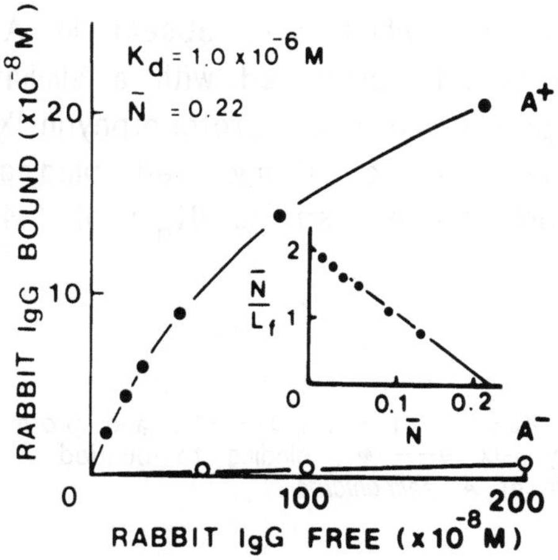

Fig.2. Kinetics of binding ^{125}I-rabbit IgG to *A. salmonicida* cells. Binding stoichiometry N were determined from Scatchard plots (inset). Binding of ^{125}I-rabbit IgG: A^+-A450, ●—● ; A^--A450-3, o—o

Neither isogenic A^- strains nor an *Aeromonas hydrophila* strain also possessing an S-layer had binding activity. Selective removal of A-protein at pH 2.2 also inactivated IgG binding. Structurally intact IgG was requisite for binding, since both Fab and Fc fragment were virtually inactive. *A. salmonicida* A-protein does not share the same IgG binding site as *Staphylococcus aureus* protein A, the archetype Fc receptor protein. Once purified, A-protein could bind IgG only weakly, but when reassembled, A-layer regained binding activity. Chemical modification or perturbation of the A-layer by a variety of reagents indicated that no single amino acid residue was critical for immunoglobulin binding, and that the binding site consisted of a native arrangement of at least four A-protein monomers in the layer. Thus the A-layer is viewed by us as a three-dimensional docking pit comprising four monomers.

Although the functional significance of immunoglobulin docking has not been demonstrated it is tempting to speculate that bound immunoglobulins (approximately 1 x 10^5/cell) act as a mechanism to shield the bacterium from host immune defenses (Munn et al. 1982), and/or help the organism enter macrophages by a nonspecific opsonization (Trust et al. 1983). Immunoglobulin binding by several Gram-positive species is well known (Langone 1982), but the functional significance has also not been elucidated.

Porphyrin Binding

Virulent A-layer containing strains of *A. salmonicida* can be conveniently detected by the ability of A^+ cells to bind the aromatic sulphonated diazo dye, Congo red, during growth on solid media (Ishiguro et al. 1981, 1985). Congo red was apparently bound by A^+ cells by a kinetically distinct mechanism (K_d, 0.25 μM) which was absent in A^- isogenic strains. Purified A-protein also bound Congo red with a similar affinity. The binding was structurally specific however, protoporphyrin-IX and hemin were strong competitive inhibitors of Congo red binding. Protoporphyrin and hemin were bound only by A^+ strains (K_d's of 0.41 and 0.63 μM, respectively) (Fig.3).

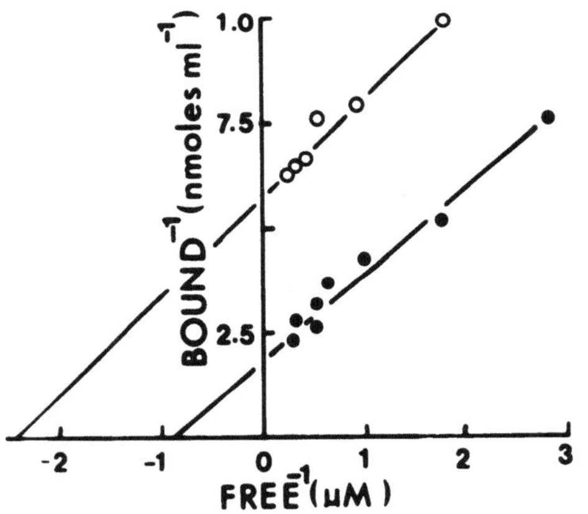

Fig.3. Kinetics of hemin, o——o, and protoporphyrin-IX, •——•, binding to purified A-protein of *A. salmonicida*

Purified A-protein also bound protoporphyrin-IX and hemin with affinities similar to those of A^+ cells (K_d's of 0.94 and 0.41 μM, respectively) (Kay et al. 1985). Thus, unlike the immunoglobulins, there appears to be a specific one-to-one high affinity association of porphyrins with A-protein monomers. This would endow each *A. salmonicida* cell with the ability to bind over half a million porphyrin molecules. The magnitude, affinity, and specificity of porphyrin binding is undoubtedly essential to *A. salmonicida.* However, there is no direct evidence that porphyrin binding is a pathogenic factor beyond the fact that A^- strains are avirulent. It is tempting to speculate that such binding represents an initial stage in iron transport. Avirulent *Shigella, Vibrio cholerae, Escherichia coli,* and *Neisseria meningitidis* mutants which have lost the ability to bind Congo red have an enhanced virulence upon supplementation with iron (Payne and Finklestein 1977). Typical *A. salmonicida* strains have two mechanisms for incorporating iron (Chart and Trust 1983), one siderophore mediated and the other requiring contact. It is possible that *A. salmonicida* has a unique heme scavenging system from their host during invasion in order to acquire iron. Some *A. salmonicida* strains are natural heme auxotrophs (Ishiguro et al. 1986), suggesting this may be an initial route of heme acquisition. *A. salmonicida* is a highly invasive pathogen. Another invasive pathogen, *Shigella flexneri,* binds Congo red and porphyrins in a manner similar to that of *A. salmonicida. S. flexneri* cells with pre-bound hemin exhibited increased invasiveness suggesting that bound porphyrins facilitate invasion (Daskaleros and Payne 1987).

Thus, the A-layer of *A. salmonicida* has so far exhibited a remarkable plethora of vital or potentially vital pathogenic functions. It appears to provide a battery of resistance strategies to thwart the combativeness of its host. Perhaps it is the summation of the properties of anti-complement activity, macrophage association and resistance, immunoglobulin binding, and porphyrin binding, which make *A. salmonicida* such a successful pathogen and explain why it has so tenaciously resisted the effective development of vaccines (Munro 1984).

References

Baumeister W, Engelhardt H (1987) Three dimensional structure of bacterial surface layers. In: Harris R, Horne RW (eds) Electron microscopy of proteins, vol 6. Academic Press, London New York

Belland RJ, Trust TJ (1985) Synthesis, export, and assembly of *Aeromonas salmonicida* A-layer analyzed by transposon mutagenesis. J Bacteriol 163:877-881

Chart H, Trust TJ (1983) Acquisition of iron by *Aeromonas salmonicida.* J Bacteriol 156:758-764

Daskaleros PA, Payne SM (1987) Congo red binding phenotype is associated with hemin binding and increased infectivity of *Shigella flexneri* in the HeLa cell model. Infect Immun 55:1393-1398

Frank MF (1979) The complement system in host defense and inflammation. Rev Infect Dis 1:483-501

Herman RL (1968) Fish furunculosis 1952-1966. Trans Am Fish Soc 97:221-230

Ishiguro EE, Kay WW, Ainsworth T, Chamberlain JB, Austen RA, Buckley JT, Trust TJ (1981) Loss of virulence during culture of *Aeromonas salmonicida* at high temperature. J Bacteriol 148:333-340

Ishiguro EE, Kay WW, Trust TJ (1984) Requirement for lipopolysaccharide in the assembly of a surface protein array involved in virulence in *Aeromonas salmonicida.* Abstr Annu Meet Am Soc Microbiol 1984 D24, p 55

Ishiguro EE, Ainsworth T, Trust TJ, Kay WW (1985) Congo red agar, a differential medium for *Aeromonas salmonicida* strains, detects the presence of the cell surface array involved in virulence. J Bacteriol 164:1233-1237

Ishiguro EE, Ainsworth T, Kay WW, Trust TJ (1986) Heme requirements for growth of fastidious atypical strains of *Aeromonas salmonicida.* Appl Environ Microbiol 51:668-670

Kay WW, Buckley JT, Ishiguro EE, Phipps BM, Monette JPL, Trust TJ (1981) Purification and disposition of a surface protein associated with virulence of *Aeromonas salmonicida.* J Bacteriol 147:1077-1084

Kay WW, Phipps BM, Ishiguro EE, Trust TJ (1985) Porphyrin binding by the surface array virulence protein of *Aeromonas salmonicida.* J Bacteriol 164:1332-1336

Langone II (1982) Protein A of *Staphylococcus aureus* and related immunoglobulin receptors produced by streptococci and pneumonococci. Adv Immunol 32:157-252

Munn CB, Ishiguro EE, Kay WW, Trust TJ (1982) Role of surface components in serum resistance of virulent *Aeromonas salmonicida.* Infect Immun 36:1069-1075

Munro ALS (1984) A furunculosis vaccine - illusion or achievable objective? In: Kinkelin P de (ed) Symp Fish Vaccination. Off Int Epizooties, Paris, pp 97-112

Payne SM, Finklestein RA (1977) Detection and differentiation of iron-responsive avirulent mutants on Congo red agar. Infect Immun 18:94-98

Phipps BM, Trust TJ, Ishiguro EE, Kay WW (1983) Purification and characterization of the cell surface virulence A-protein from *Aeromonas salmonicida.* Biochemistry 22:2934-2939

Sleytr UB, Messner P (1983) Crystalline surface layers on bacteria. Annu Rev Microbiol 37:311-339

Trust TJ, Kay WW, Ishiguro EE (1983) Cell surface hydrophobicity and macrophage association of *Aeromonas salmonicida.* Curr Microbiol 9:315-318

Udey LR, Fryer JL (1978) Immunization of fish with bacterins of *Aeromonas salmonicida.* Mar Fish Rev 40:12-17

Wrathmell AB, Parish NM (1980) Cell surface receptors in the immune response of fish. In: Manning MJ (ed) Immunological memory. Elsevier/North Holland, Amsterdam, pp 143-152

Thermoresistance of A-Layer - Deficient Mutants of *Aeromonas salmonicida*

E.E. Ishiguro

Department of Biochemistry and Microbiology
University of Victoria
Victoria, B.C., V8W 2Y2
CANADA

Aeromonas salmonicida is the causative agent of the fish disease known as furunculosis. The primary virulence factor of *A. salmonicida* is a cell surface protein array called the A-layer, and mutants (A^-) lacking the A-layer are avirulent (Udey and Fryer 1978, Ishiguro et al. 1981). The principal component of the A-layer is a 50-kDa protein (A-protein) which has been purified and characterized (Phipps et al. 1983). Evidence suggests that the A-layer represents a relatively impermeable outermost barrier which completely shields the cell wall outer membrane. The A-layer probably provides wild-type (A^+) bacteria with a protective barrier against the defense mechanisms of the fish host.

We have previously reported that a single subculture of *A. salmonicida* at high temperature results in the loss of virulence due to the growth of A^- derivatives (Ishiguro et al. 1981). In order to maintain virulence, the growth temperature must be maintained below 26 to 27^oC. These thermoselected A^- mutants apparently completely lack the ability to synthesize A-protein under all culture conditions. This loss of virulence was stable, and reversion to the A^+ state could not be detected. We have speculated that this phenomenon was due to the thermoselection of rare A^- mutants in the original population. Two experiments have been performed to gain support for this hypothesis.

The first set of experiments dealt with the origin of the A-derivatives of wild-type *A. salmonicida.* I have used the standard Luria-Delbruck fluctuation test (Luria and Delbruck 1943) to prove that these A^- derivatives arise spontaneously during the normal growth of wild-type strains (data not shown). These experiments established that A^- derivatives pre-exist at low frequencies in cultures of wild-type A^+ strains grown at low temperatures. They do not arise as a direct consequence of high temperature. For example, one popular notion was that plasmid curing induced by high temperature was involved. Assuming that these A^- derivatives arise as a result of spontaneous mutations (see below), a mutation rate of 2 to 5 x 10^{-8} was calculated from these experiments.

The second series of experiments addressed the basis for the selection of the A^- derivatives at high temperature. Figure 1 summarizes the results of an experiment in which the growth temperature ranges of a wild-type A^+ strain, A451, and a thermoselected A^- mutant derived from it, strain A451-2 (Ishiguro et al. 1981), were determined. Cultures of these strains were grown to stationary phase at low temperature (20°C). The abilities of these cells to form colonies at temperatures ranging from 20 to 32°C were determined by plating dilutions of the cultures on Congo red-TS agar plates and incubating these at the indicated temperatures. In Fig.1, the viable cell counts relative to those obtained at 20°C are plotted as a function of the incubation temperature. The results clearly establish that the maximum growth temperature of the wild-type strain was about 26°C, but was significantly higher (30°C) in the case of its A^- derivative. As noted previously, Congo red-TS agar is a useful medium for differentiating A^+ and A^- derivatives of *A. salmonicida.* A-protein avidly binds Congo red, and A^+ colonies are consequently bright red on this medium. On the other hand, exposed components of the cell wall outer membrane do not bind Congo red, and the intact outer membrane is impermeable to Congo red. Thus, colonies of A^- bacteria are virtually colourless on Congo red-TS. In this respect, the colonies formed by strain A451 were phenotypically A^+ at low temperatures. On the other hand, all of the colonies formed above 26°C were phenotypically A^-, and these therefore apparently represented the pre-existing A^- derivatives noted in the first set of experiments. Identical results were obtained with five other pairs of A^+ and A^- strains. These results are consistent with the idea that A^- -derivatives have higher growth temperature maxima than their A^+ parents and consequently are selected by incubation at high temperatures.

We (Ishiguro et al. submitted) have isolated avirulent mutants of six different A^+ strains which were unable to synthesize the O-oligosaccharide portion of the outer membrane lipopolysaccharide (LPS). Such mutants were able to synthesize A-protein normally but they released substantial amounts of A-protein into the culture medium and

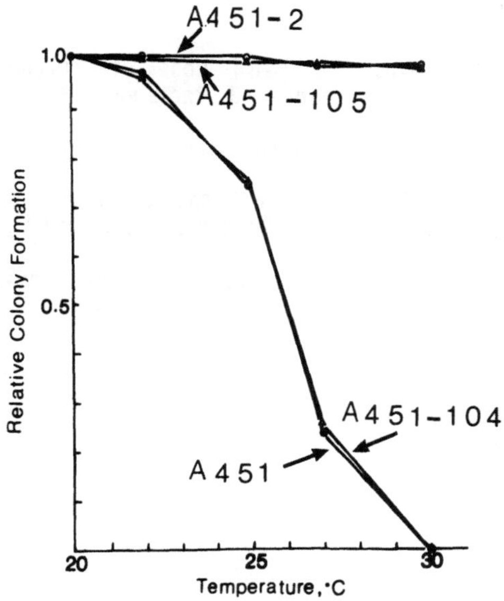

Fig.1. Growth temperature relationships of strains A451 (wild-type), A451-2 (thermoselected A$^-$ mutant), A451-104 (LPS-defective mutant), and A451-105 (A$^-$ derivative of A451-104). The relative colony-forming capabilities of each strain are shown as a function of growth temperature

were consequently phenotypically A$^-$. In every case, these mutants exhibited the temperature-sensitive growth characteristic of their A$^+$ parents, e.g., LPS mutant strain A451-104 (Fig.1). Furthermore, thermoresistant derivatives of such mutants, e.g. strain A451-105 (Fig.1), were always completely deficient in A-protein synthesis. From these results, I suggest that some aspect of A-protein synthesis may be responsible for the thermosensitivity of A$^+$ bacteria. R.J. Belland and T.J. Trust (Belland, Ph.D. Thesis, University of Victoria, 1987) have shown that the only thermoresistant A$^-$ strain examined to date carried a deletion of the 5′-end of the structural gene for A-protein. If this proves generally true for all such mutants, it would appear that even trace expression of A-protein synthesis is sufficient to inhibit growth at 30°C. In view of the magnitude of A-protein synthesis in *A. salmonicida,* I speculate that an event unique to the process, e.g. A-protein export, represents the factor which dictates the maximum growth temperature in this organism. In this respect, it is noteworthy that the so-called atypical strains of *A. salmonicida* differ from the typical strains described here in that they grow well and synthesize A-protein at 30°C.

References

Ishiguro EE, Kay WW, Ainsworth T, Buckley JT, Trust TJ (1981) Loss of virulence during culture of *Aeromonas salmonicida* at high temperature. J Bacteriol 148:333-340

Luria SE, Delbruck M (1943) Mutations of bacteria from virus sensitivity to virus resistance. Genetics 28:491-511

Phipps BM, Trust TJ, Ishiguro EE, Kay WW (1983) Purification and characterization of the cell surface virulent A-protein from *Aeromonas salmonicida.* Biochemistry 22:2934-2939

Udey LR, Fryer JL (1978) Immunization of fish with bacterins of *Aeromonas salmonicida.* Mar Fish Rev 40:12-17

The S-Layer of Pathogenic Strains of *Aeromonas hydrophila*

J.S.G. Dooley[1], R.G.E. Murray[2] and T.J. Trust[1]

[1]Department of Biochemistry and Microbiology,
University of Victoria
Victoria, British Columbia, V8W 2Y2, CANADA
[2]Department of Microbiology and Immunology
University of Western Ontario
London, Ontario, N6A 5C1, CANADA

Aeromonas hydrophila, a Gram-negative, motile rod, is a pathogen of a wide variety of animals, including man. In man the organism is most commonly implicated as a gastrointestinal pathogen, but has also been isolated from a variety of other clinical conditions (Trust and Chipman 1979). The species is especially important as a pathogen of fish, typically producing a fulminant haemorrhagic septicaemia (Trust 1986). In many instances disease most likely results from opportunistic infection by strains of *A. hydrophila* from the fishes' normal flora. In other cases, however, the strains of *A. hydrophila* producing fish disease clearly behave as primary pathogens (Mittal et al. 1980, DeFigueiredo and Plumb 1977). For example, a group of strains with high virulence for salmonids has been reported by Mittal et al. (1980). These strains exhibit certain cell-surface-associated phenotypic characteristics unique among *A. hydrophila* isolates. They auto-aggregate in static broth culture, are resistant to the bactericidal activity of normal serum, and form a single thermostable serogroup based on the O-polysaccharide antigens of their lipopolysaccharides (LPS) (Mittal et al. 1980, Dooley et al. 1985). Several members of the serogroup, while exhibiting virulence for fish, were not isolated from fish, but were isolated from sources such as aborted piglet liver, bovine brain, and human diarrhoeal faeces.

Electron microscopy revealed the presence of a tetragonally arrayed S-layer on all members of this high virulence *Aeromonas hydrophila* serogroup. The S-layers were similar on all the strains examined, with unit cell dimensions of approximately 12 nm. A single representative strain, strain TF7, was selected for further analysis. Freeze-cleaved and etched preparations and sections for electron microscopy showed that the S-layer was the outermost component of the cell envelope. This was confirmed by observation of thin sections. Both pili and flagella were also found in this strain. Computer-generated enhancements of the negatively stained micrographs showed the subunit organization to a resolution of <4 nm. Two structural units of identical lattice constants alternated in the array in both axes and one of them was apparently dominant as the centre of mass. The lesser unit was rotated 20° from the dominant axis of symmetry and was formed by the junction of linker projections from a corner of the four components of the dominant unit. The unit cell as defined showed p4 symmetry, and a = b = 12.2 nm.

The morphological interpretation was supported by finding that the array consists of a single polypeptide of apparent molecular weight 52,000 (M_r 52 K). Treatment of whole cells of *A. hydrophila* strain TF7 with 0.2 M glycine buffer pH 4.0 resulted in the release of sheets of S-layer material, and sodium dodecyl sulphate (SDS)-polyacrylamide gel electrophoretic (SDS-PAGE) analysis showed that this sheet material was composed primarily of a protein of M_r 52 K. A M_r 52 K protein was also the predominant protein in glycine extracts of other members of the high virulence serogroup. Immunoblotting with antiserum raised against formalinized whole cells of *A. hydrophila* TF7 showed the 52K S-layer protein to be the major surface protein antigen (Dooley et al. 1986), and impermeant Sulfo-NHS-Biotin cell surface labelling (Hurley et al. 1985) showed that the 52K S-layer protein was the only protein accessible to the Sulfo-NHS-Biotin label, and effectively masked underlying outer membrane (OM) proteins. In its native surface conformation the 52K S-layer protein was only weakly radiolabelled by a lactoperoxidase ^{125}I-surface radiolabelling procedure.

The other major surface antigen of the high virulence group of *A. hydrophila* was the O-polysaccharide of the lipopolysaccharide (LPS). SDS-PAGE analysis showed that these O-polysaccharides were of homogeneous chain length, in contrast to the heterogeneous chain length O-polysaccharides of other *A. hydrophila* strains (Dooley et al. 1985). Immunofluorescence microscopy showed that a number of these O-polysaccharide chains penetrated the S-layer, and were exposed on the cell surface. A UV-induced rough lipopolysaccharide mutant of TF7 was found to produce an intact S-layer, but a deep rough LPS mutant was unable to assemble an array and excreted the S-layer protein into the growth medium, indicating that a minimum LPS oligosaccharide size is required for the assembly of a typical S-layer on the *A. hydrophila* cell surface. The 52K S-layer protein exhibited heat-modifiable SDS-solubilization behaviour when associated with OM, but was fully solubilized by SDS at all temperatures after removal from the OM, indicating a strong interaction of S-layer with underlying OM. The native S-layer was permeable to ^{125}I in the lactoperoxidase radiolabelling procedure, and two major OM proteins of M_r 30K and 48K were iodinated. The M_r 48K species was a peptidoglycan-associated, transmembrane protein which exhibited heat modifiable SDS-solubilization behaviour characteristic of a pore protein (Hancock 1987). A 50K major peptidoglycan-associated OM protein which was not radiolabelled exhibited similar SDS-heat solubilization characteristics, and possibly represents a second pore protein.

The S-layer protein of *A. hydrophila* TF7 was extracted from whole cells by treatment with 0.2 M glycine-HCl pH 3.0, and the subunit protein was purified to homogeneity by ion exchange chromatography and reverse phase high performance liquid chromatography. Amino acid composition analysis showed that the protein contained 520 residues per molecule, 41% of which were hydrophobic. Cysteine was absent. A pI of 4.6 was determined for the protein and only a single isoelectric form was detected. The purified protein displayed the hydrophobic characteristic of binding to Octyl-Sepharose gels, but the salt aggregation test (Lindahl et al. 1981) showed that it did not confer hydrophobicity to the cell surface when present as an intact S-layer. The molecule aggregated strongly in aqueous solution as determined by sedimentation equilibrium studies. Circular dichroism spectra showed that the S-layer protein was composed of a large amount of β-sheet (approximately 44%), a limited amount of α-helix (19%), and 12% β-turn with the remainder of the molecule aperiodic in nature. No difference in secondary structure content was measured in the presence of the metal chelator EDTA. The N-terminal amino acid sequence was determined for the first 30 residues. Little or no sequence homology was shown with the S-layer protein purified from another strain of *A. hydrophila,* with the S-layer proteins of *Aeromonas salmonicida* (Phipps et al. 1983), and *Campylobacter fetus* (Dubreuil and Trust unpubl.), the glycoprotein S-layer protein of *Halobacterium halobium* (Lechner and Sumper 1987), or the outer wall protein of *Bacillus brevis* 47 (Tsuboi et al. 1986).

The S-layer proteins of various strains of *A. hydrophila* differ antigenically, and these antigenic differences appear to reside in surface-exposed epitopes. This contrasts to the high degree of antigenic conservation of the *A. salmonicida* surface array proteins (Kay et al. 1984). The *A. hydrophila* S-layer also differs from the *A. salmonicida* surface array in several biological functions (Ishiguro et al. 1985, Trust et al. 1983), appearing not to bind Congo red, or to facilitate preferential association with macrophages. This is most likely a reflection of the different modes of pathogenesis of the two species of *Aeromonas,* with *A. hydrophila* being far more dependent on toxins to produce disease. However, as is the case in *A. salmonicida* (Munn et al. 1982), the *A. hydrophila* array most likely facilitates initial colonization of host tissues by contributing to the ability of the organism to resist the bactericidal activity of non-immune serum.

Acknowledgements.
This study was supported in part by grants from the Medical Research Council of Canada to RGEM, and the Natural Sciences and Engineering Research Council of Canada, and the British Columbia Health Care Research Foundation to TJT. We are grateful to Mrs. M. Hall for expert technical assistance, Dr. P.W. Whippey for image enhancement, and Drs. W.D. Mc Cubbin and C.M. Kay for sedimentation equilibrium and circular dichroism analyses.

References

Dooley JSG, Lallier R, Shaw DH, Trust TJ (1985) Electrophoretic and immunochemical analyses of the lipopolysaccharides from various strains of *Aeromonas hydrophila.* J Bacteriol 164:263-269

Dooley JSG, Lallier R, Trust TJ (1986) Antigenic structure of *Aeromonas hydrophila.* Vet Immunol Immunopathol 12:339-344

Figueiredo J De, Plumb JA (1977) Virulence of different isolates of *Aeromonas hydrophila* in channel catfish. Aquaculture 11:349-354

Hancock REW (1987) Role of porins in outer membrane permeability. J Bacteriol 169:929-933

Hurley WL, Finkelstein E, Holst BD (1985) Identification of surface proteins on bovine leukocytes by a biotin-avidin protein blotting technique. J Immunol Methods 85:195-202

Ishiguro EE, Ainsworth T, Trust TJ, Kay WW (1985) Congo red agar, a differential medium for *Aeromonas salmonicida,* detects the presence of the cell surface protein array involved in virulence. J Bacteriol 164:1233-1237

Kay WW, Phipps BM, Ishiguro EE, Olafson RW, Trust TJ (1984) Surface layer virulence A-proteins from *Aeromonas salmonicida* strains. Can J Biochem Cell Biol 62:1064-1071

Lechner J, Sumper M (1987) The primary structure of a procaryotic glycoprotein. J Biol Chem 262:9724-9729

Lindahl M, Faris A, Wadstrom T, Hjerten S (1981) A new test based on salting out to measure relative surface hydrophobicity of bacterial cells. Biochim Biophys Acta 677:471-476

Mittal KR, Lalonde G, Leblanc D, Olivier G, Lallier R (1980) *Aeromonas hydrophila* in rainbow trout: relation between virulence and surface characteristics. Can J Microbiol 26:1501-1503

Munn CB, Ishiguro EE, Kay WW, Trust TJ (1982) Role of surface components in serum resistance of virulent *Aeromonas salmonicida*. Infect Immun 36:1069-1075

Phipps BM, Trust TJ, Ishiguro EE, Kay WW (1983) Purification and characterization of the cell surface virulence A-protein from *Aeromonas salmonicida*. Biochemistry 22:2934-2939

Trust TJ (1986) Pathogenesis of infectious diseases of fish. Annu Rev Microbiol 40:479-502

Trust TJ, Chipman DC (1979) Clinical involvement of *Aeromonas hydrophila*. J Can Med Assoc 120:942-947

Trust TJ, Kay WW, Ishiguro EE (1983) Cell surface hydrophobicity and macrophage association of *Aeromonas salmonicida*. Curr Microbiol 9:315-318

Tsuboi A, Uchihi R, Tabata R, Takahashi Y, Hashiba H, Sasaki T, Yamagata H, Tsukagoshi N, Udaka S (1986) Characterization of the genes coding for two major cell wall proteins from protein-producing *Bacillus brevis* 47: complete nucleotide sequence of the outer wall protein gene. J Bacteriol 168:365-373

The Three-Dimensional Structure of Bacterial Surface Layers

S. Hovmöller, A. Sjögren and D.N. Wang

Structural Chemistry
University of Stockholm
S-106 91 Stockholm
SWEDEN

The research of bacterial surface layers (S-layers) is in the unusual situation that more is known about the structure than about the function of S-layers. While it has been argued that we must first understand the function of S-layers before there is any point in studying their structures, we believe structure determinations to be important today. A more detailed knowledge of the three-dimensional structures of S-layers may help in suggesting experiments for finding their functions.

Electron Microscopy

Electron microscopy has been the main tool for studying S-layer structure, ever since S-layers were first found by this technique. The S-layers are very well suited for electron microscopy, and especially for transmission electron microscopy (TEM). Because of the strong interaction of electrons with matter, only very thin objects (<30 nm thick) can be studied by TEM. S-layers are formed as a single layer of protein molecules, and the thickness of this layer is usually 5 to 20 nm. The crystalline order of S-layers is also a very important advantage for structure determination. Because of radiation damage and non-uniform staining of individual protein molecules, the signal-to-noise ration for images of single protein molecules is low. If the molecules are ordered in one-, two- or three-dimensional arrays it is possible to make averages over many molecules, each exposed less to the electron irradiation, thus obtaining a reliable rather detailed structure.

Image Processing

The averaging of structure information from an ordered array of molecules is performed by computer. First the electron micrograph is digitized, for example into 512 x 512 points, each 40 x 40 micrometres.

In this way an area of 2 x 2 cm^2 is digitized. If only a smaller area on the micrograph shows a good order, then a smaller area can be used for the study. If the magnification used is 40,000 times, then each sample point represents an area of 10 x 10 $Å^2$ of the S-layer. Such a sampling is sufficiently fine to restore all information present in standard negatively stained preparations, which will typically be limited to about 25 Å resolution (Fig.1). If the unit cell dimensions of the S-layer are 15 nm, then each unit cell is sampled at 15 x 15 = 225 points and about 1000 unit cells are digitized. The digitized images are stored on computer. The averaging may be carried out by Fourier transform techniques, or by correlation averaging or by a combination of the two. When the S-layer crystals are well ordered the methods are equivalent, but only correlation techniques are able to compensate for bent lattices etc. In our studies only Fourier techniques were used.

The Fourier transform of the digitized area is calculated by computer. This transform contains all the information of the scanned image, but the information is organized in a different way. In the original image (so-called real space) each individual protein molecule with its noise can be seen. In the Fourier transform (so-called reciprocal space) the periodic features of the image are emphasized. The Fourier transform is a complex function (A + iB) of the same size as the image scanned,

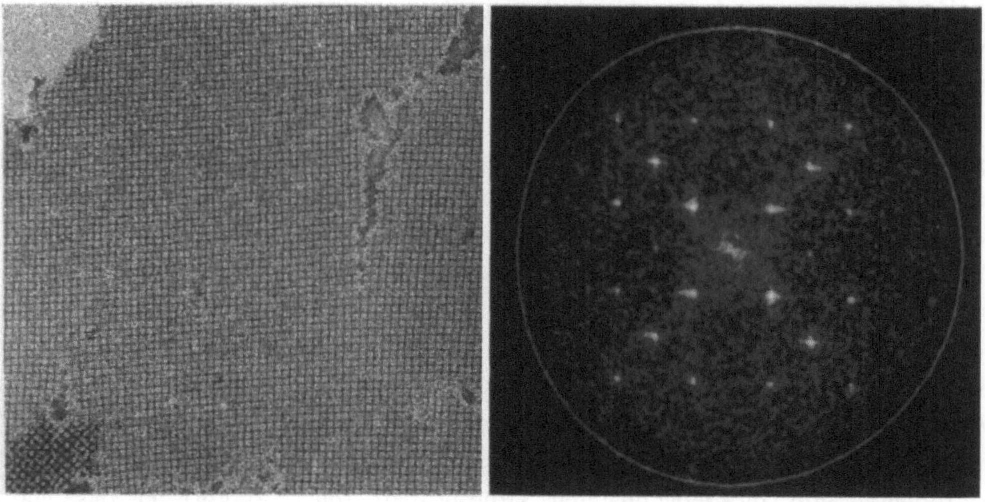

Fig.1. (A) Electron micrograph of the S-layer from *Eubacterium* sp ES4C. While a large portion of this layers shows perfect order, about one third can be seen to have a slightly different orientation. In the corner a Moire pattern can be seen as a result of two superimposed layers. (B) The diffraction pattern obtained from a perfectly ordered area of the same S-layer. The tetragonal (fourfold) symmetry can be seen both in the image and in the diffraction pattern. The ring indicates 25 Å resolution

typically 512 x 512 points. The amplitude part of the transform is simply calculated as $A^2 + B^2$ for each point, and displayed on the computer graphics screen. This is the diffraction pattern of the crystal. From the diffraction pattern the symmetry and unit cell dimensions of the S-layer can immediately be determined. The distance between diffraction spots is the reciprocal of the unit cell dimensions (hence the name reciprocal space). Each spot in the diffraction pattern shows the presence of a set of parallel lines. The closer the lines are together, the further the spot is from the centre of the diffraction pattern. The fine details, i.e. the high resolution information, is therefore to be found in the outermost diffraction points. The better the crystal, the more diffraction points and the higher resolution can be obtained. The position of each diffraction point corresponds to the normal of the set of parallel lines in the image of the crystal. The stronger the lines are, the stronger is the diffraction spot, its amplitude.

The non-periodic features of the image, the noise, can be filtered out from the image by computer processing. This is possible because the noise is evenly spread over the whole of reciprocal space, and not limited to the exact lattice positions as the information from the periodic crystal is. When the inverse Fourier transform is calculated, only the information on the lattice is used, and the result is a virtually noise-free averaged image of the protein molecules of the S-layer. It may be emphasized that this procedure is possible starting from electron micrographs, but not from X-ray diffraction patterns. The reasons for this is that in X-ray diffraction only the above-mentioned information is present, but in the computer-calculated Fourier transform of an image also the <u>phase</u> information is restored. The phase of a diffraction point describes the relative <u>position</u> of the set of parallel lines in the image, and this is needed in the image reconstruction.

The S-layer structures may be further improved by imposing the exact crystal symmetry, and by merging data from several images. Finally it is also possible to make a three-dimensional determination of the structure by adding a series of images taken from different directions of view, using a tilting stage in the electron microscope. We have determined the three-dimensional structure of four S-layers in this way. They are *Synechocystis* CL II, hexagonal p6 (Karlsson et al. 1983), *Bacteroides buccae* , hexagonal p6 (Sjögren et al. 1985) and two species of *Eubacterium* (Fig.2).

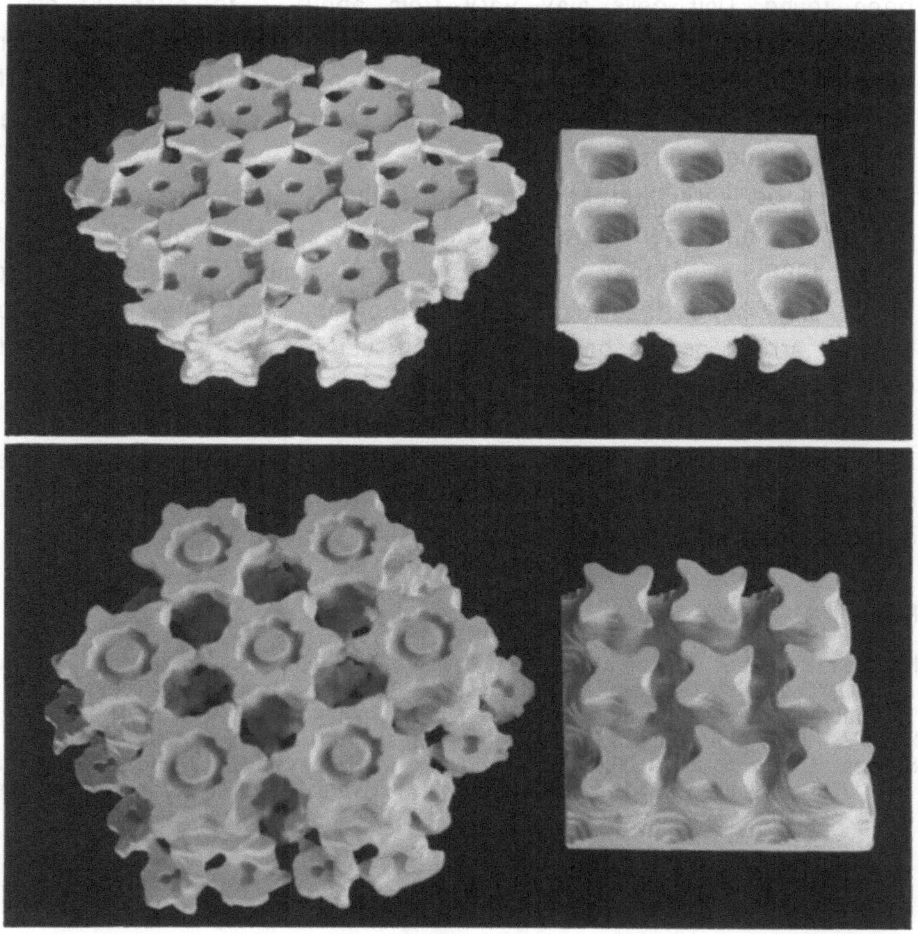

Fig.2. Three-dimensional structures of the S-layers of *Eubacterium* sp. AHN 990 (left) and *Eubacterium* sp ES4C with p6 and p4 symmetry, respectively

Comparison of S-Layer Structures

With the number of 3-D structures of S-layers now being about 20, we can start to make some general comments on these structures. The S-layers are made of protein monomers, often first aggregating into oligomers with 2, 3, 4 or 6 monomers. These then pack into a rather well-ordered crystal lattice, covering the entire bacterial cell. The most common symmetry is hexagonal, denoted p6, followed by tetragonal p4. Of the other possible symmetries (p1, p2 and p3), only p1 has not yet been found. Unit cells may vary from about 7 to over 20 nm in the plane of the crystal, and the thicknesses are from some 5 to 20 nm. The variation of structures even within a given symmetry and unit cell size is unlimited, and the S-layers studied so far show a wide variety of packing arrangements.

It will be very interesting to see if the S-layers are generally closely related, as has been shown for the surface proteins of spherical viruses (Liljas 1986) or if they represent a large number of unrelated proteins, which only have that one thing in common that they aggregate into ordered arrays on the bacterial surface just because they are secreted in such a large quantity.

The answers to these exciting questions may come in the next 5 or 10 years, from the protein sequence data that has just started to accumulate, and from higher resolution (atomic resolution) structure determinations of S-layers.

References

Karlsson B, Vaara T, Lounatmaa K, Gyllenberg H (1983) Three-dimensional structure of the regularly constructed surface-layer (S-layer) from *Synechocystis* CL II. J Bacteriol 156:1338-1343

Liljas A (1986) The structure of spherical viruses. Prog Biophys Mol Biol 48:1-36

Sjögren A, Hovmöller S, Farrants G, Ranta H, Haapasalo M, Ranta K, Lounatmaa K (1985) Structures of two different surface layers found in six *Bacteroides* strains. J Bacteriol 164:1278-1282

Characterization of the Surface Protein of *Comamonas acidovorans*

S. Gerbl-Rieger, S. Schneider, H. Engelhardt, J. Peters and W. Baumeister

Max-Planck-Institut für Biochemie
D-8033 Martinsried bei München
FRG

Comamonas acidovorans, formerly *Pseudomonas acidovorans* (ATCC 15668) (Tamaoka et al. 1987), possesses a regularly arrayed surface protein (Fig.1) which is packed in a simple "cobblestone" fashion. The p4 lattice has a centre-to-centre spacing of 11 nm (Chalcroft et al. 1986). This structural motif occurs in several species of the β-subdivision of purple bacteria (Fig.2) (Acker et al. 1975, Beveridge and Murray 1975, 1976, Lapchine 1979, Wells et al. 1983).

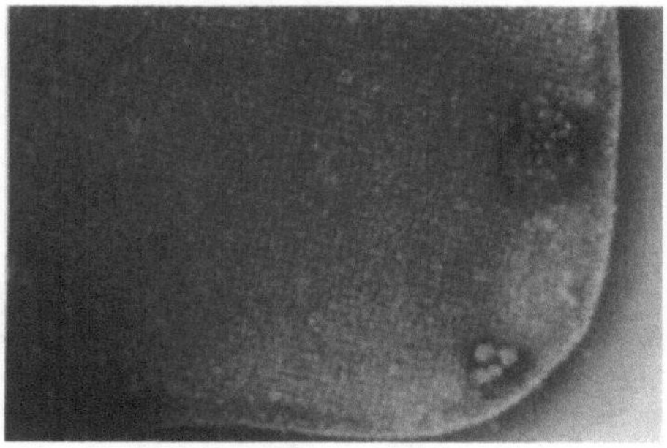

Fig.1. Negatively stained outer membrane of *Comamonas acidovorans* showing a tetragonal lattice

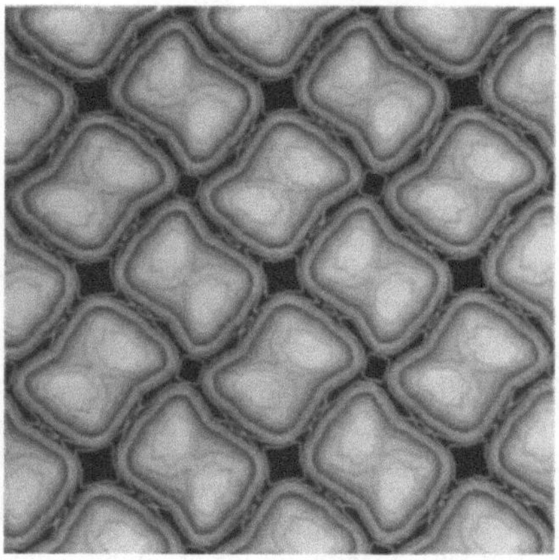

Fig.2. Two-dimensional average of the *Comamonas acidovorans* outer membrane including a fourfold symmetrization

While all attempts to separate the regular protein layer from the outer membrane failed, solubilization with SDS invariably results in the disintegration of the protein lattice. Thus, the protein appears to be an integral component of the outer membrane (a "rOmp"), but a substantial portion of it is exposed on the outer surface. When the solubilized outer membrane components were analyzed on an SDS-polyacrylamid gel several high apparent molecular weight bands appeared along with bands at apparent M_r of 32,000 and 21,000 (Fig.3).

Treatment of the samples with BrCN in 70% formic acid before running the SDS-PAGE (Fig.4) led to the disappearence of all high molecular weight bands with a concomitant increase of intensity of the band at M_r 32,000. Thus, the high molecular weight bands represent oligomeric complexes or aggregates of the M_r 32,000 protein which must be regarded as the major (if not the only) outer membrane protein component.

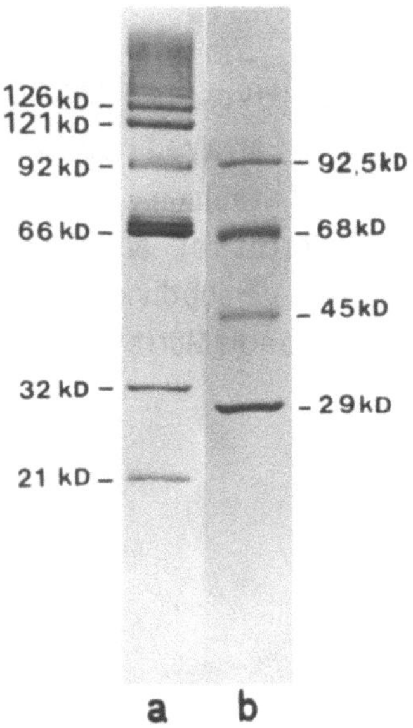

Fig.3. The polypeptide pattern of the outer membrane fraction of *Comamonas acidovorans.* SDS-PAGE: 5-16% polyacrylamid gradient gel. Lane a: The outer membrane polypeptides were isolated by a Percoll density gradient centrifugation, treated with Triton X-100, solubilized with SDS and reduced with Dithiothreitol. Lane b: M_r standard-proteins

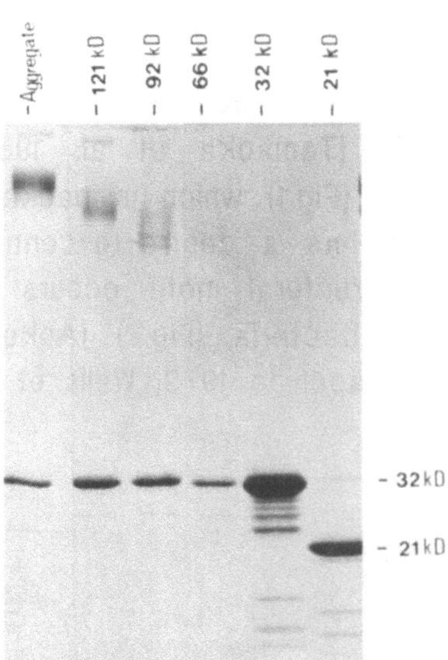

Fig.4. Treatment of the isolated M_r 32,000 and M_r 21,000 polypeptides with BrCN in 70% formic acid. Fig.3 shows that the high M_r aggregates are oligomeric forms of the M_r 32,000 monomer. SDS-PAGE: 10-16%

Both the M_r 32,000 and M_r 21,000 polypeptides were only partially cleaved by BrCN. A comparison of the cleavage patterns indicates that the M_r 21,000 polypeptide is not a proteolytic fragment of the M_r 32,000 polypeptide. We have determined the stoichiometric ratio of the M_r 32,000 and Mr 21,000 proteins by comparing peak areas (calculated from quantitative band shape analyses) of HPLC runs of several preparations of the outer membrane; the composition of each fraction was checked by SDS-PAGE (Fig.5a,b). The specific absorbance coefficients were estimated using quantitation by amino acid analysis. The average molar ratio of the apparent M_r 21,000 protein (C) over the apparent M_r 32,000 protein (B) including the aggregates (A) turned out to be 1 : 9 (Fig.6). This ratio, in conjunction with symmetry considerations (see also below), makes it very unlikely that the M_r 21,000 protein is an integral component of the regular lattice. Its location in the cell envelope remains to be elucidated.

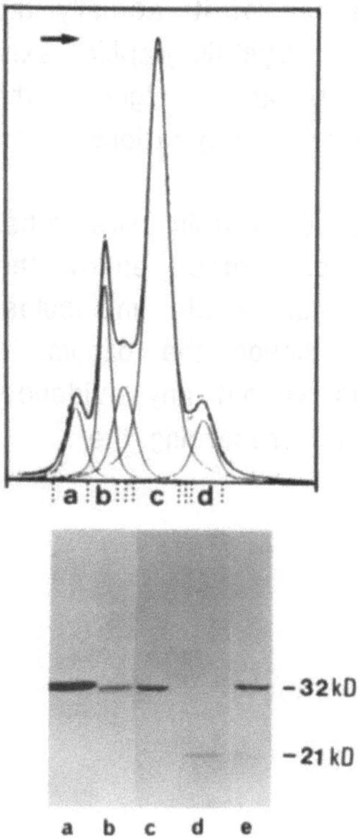

Sample / Protein in µg	total peak area	peak area of aggregates A	peak area of 32 Kdal protein B	peak area of 21 Kdal protein C	C/A+B
P_2-024 / 750	36,38	13,59	19,90	2,88	1:7,62
P_2-029 / 450	29,86	10,64	17,31	1,63	1:11,25
P_3-039 / 715	94,76	42,74	42,16	7,16	1:7,70
P_3-044 / 28,6	23,92	7,14	14,17	1,54	1:9,08
P_5-045 / 0,375	21,75	6,44	13,84	1,44	1:9,24
				Ø	1:8,98

Fig.5. (a) Separation of the outer membrane polypeptides of *Comamonas acidovorans* by molecular sieve chromatography on a TSK 3000 column (detection at 206 nm). 0.1% TFA/30% acetonitrile was used as solvent. ⟶ : retention time. The fractions a,b,c,d were collected and separated by SDS-PAGE after a treatment with 70% formic acid. (b) Polypeptide pattern of the fractions a,b,c,d. SDS-PAGE: 5-18% polyacrylamide. Lane a: fraction a; lane b: fraction b; lane c: fraction c; lane d: fraction d; lane e: outer membrane polypeptides not separated

Fig.6. Quantitative evaluation of the stoichiometric ratio of the M_r 21,000 and the M_r 32,000 proteins. The molar ratio was investigated by comparing peak areas of several protein preparations separated by HPLC (Fig.5a)

We have performed STEM mass determination (Engel 1978) in order to relate the apparent M_r of the rOmp to its structure as revealed by electron microscopy and image processing. STEM mass determination (kindly performed by A. Engel, Basel) yielded a mass of 625,000 40,000 per unit cell. Supposing a lipid-protein ratio of the membrane of 20 : 80 (calculated from infraredspectra of the outer membrane) the molecular weight of one morphological complex (one-half of the unit cell; Fig.2) is about 270,000. Analytical equilibrium density gradient ultracentrifugation (kindly performed by J. Rosenbusch, Basel) yielded values very similar to the STEM measurements. Two bands were found. One corresponds to the M_r of 270,000 and probably represents the whole complex. The band at M_r of 140,000 is close to that expected for one half of it; actually the complex shows a faint division near the twofold crystallographic axis (Fig.2); i.e. it could be composed of two tetramers (given the unreliability of apparent M_r from SDS-PAGE) other configurations such as two trimers are also possible.

It is reasonable to assume that the M_r 32,000 protein, like other outer membrane proteins (Benz 1975), in one form or another, endow the membrane with the necessary permeability for small solute molecules. Hence it may be referred to as a porin even though the oligomeric complex appears monolithic on electron micrographs without any evidence of a pore at the present resolution of 2 nm. Without disregarding the

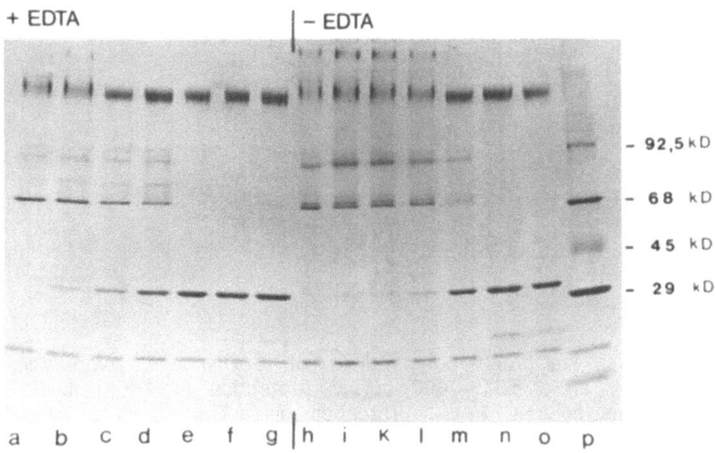

Fig.7. Influence of solubilization temperature and EDTA-treatment on the outer membrane polypeptide pattern of *Comamonas acidovorans*. SDS-PAGE: 5-16% polyacrylamide. Temperatures: lane a,h: 50 $^{\circ}$C; lane b,i: 60 $^{\circ}$C; lane c,k: 65 $^{\circ}$C; lane d,l: 70 $^{\circ}$C; lane e,m: 80 $^{\circ}$C; lane f,n: 90 $^{\circ}$C; lane g,o: 100 $^{\circ}$C; lane p: M_r-standard proteins

possibility of a pore emerging at a higher resolution, one could also imagine pores to be formed between the complexes assuming the protein to organize the surrounding lipid in a peculiar fashion. Actually the M_r 32,000 protein shares some properties with bona fide porins (such as Omp F from *E. coli* (Kleffel et al. 1985), e.g. it has a -structure content of about 65% (data not shown). The effect of temperature and EDTA on the aggregational state of the M_r 32,000 polypeptide was studied by SDS-PAGE (Fig.7). At temperatures above 70°C oligomers dissociate into monomers of apparent M_r 32,000. A similar behaviour was found with the major Omp of *Rhodopseudomonas capsulata* (Flamman and Weckesser 1984) and with the Omp F of *E. coli* K12 (Nakamura and Mizushima 1976).

We are currently engaged in cloning and sequencing the genes of the outer membrane proteins, which will enable us to search for pertinent homologies. One of the attractions of this membrane is its rather simple composition and basic architecture offering a good chance to obtain a detailed and yet a "holistic" model of a bacterial outer membrane.

References

Acker G, Schmitt R, Mayer F (1975) Elektronenmikroskopische Untersuchungen alternder Zellen von *Pseudomonas rhodos.* Arch Microbiol 104:215-233

Benz R (1975) Porin from bacterial and mitochondrial outer membranes. CRC Crit Rev Biochem 19:145-190

Beveridge TJ, Murray RGE (1975) Surface arrays on the cell wall of *Spirillum metamorphum.* J Bacteriol 124:1529-1544

Beveridge TJ, Murray RGE (1976) Reassembly in vitro of the superficial cell wall components of *Spirillum putridiconchylium.* J Ultrastruct Res 55:105-118

Chalcroft J, Engelhardt H, Baumeister W (1986) Three-dimensional structure of a regular surface layer from *Pseudomonas acidovorans.* Arch Microbiol 144:196-200

Engel A (1978) Molecular weight determination by scanning transmission electron microscopy. Ultramicroscopy 3:273-281

Flamman HT, Weckesser J (1984) Characterization of the cell wall and outer membrane of *Rhodopseudomonas capsulata.* J Bacteriol 159:191-198

Kleffel B, Garavito RH, Baumeister W, Rosenbusch JP (1985) Secondary structure of a channel-forming protein: porin from *E. coli* outer membranes. EMBO J 4:1589-1592

Lapchine L (1979) Regularly arranged structures on the surface of some *Pseudomonas* sp.. FEMS Microbiol Lett 5:223-225

Nakamura K, Mizushima S (1976) Effects of heating in dodecyl sulfate solution on the conformation and electrophoretic mobility of isolated major outer membrane proteins from *Escherichia coli* K-12. J Biochem 80:1411-1422

Tamaoka J, Ha DH, Komagata K (1987) Reclassification of *Pseudomonas acidovorans* den Dooren de Jong 1926 and *Pseudomonas testosteroni* Marcus and Talalay 1956 as *Comamonas acidovorans* comb. nov. and *Comamonas testosteroni* comb. nov., with emended description of the genus *Comamonas.* Int J Syst Bacteriol 37:52-59

Wells B, Horne RW, Lund BH, King NR (1983) The ultrastructure of *Pseudomonas avenae* - I.Paracrystalline surface layer and extracellular material. Micron 14:11-28

The Surface Protein of the Archaebacterium *Thermoproteus tenax*

I. Wildhaber[1], A. Engel[2] and W. Baumeister[1]

[1]Max-Planck-Institut für Biochemie
D-8033 Martinsried bei München, FRG
[2]Maurice E. Müller-Institut
Biozentrum der Universität Basel
Klingelbergstr. 70
CH-4056 Basel, SWITZERLAND

The Thermoproteales are extremely thermophilic, anaerobic and sulphur-dependent archaebacteria living in habitats reminscent of the environmental conditions at early stages in the biotic evolution.

The three-dimensional structure of the surface protein of two species of the Thermoproteales has been investigated by electron microscopy: the surface layer of *Thermoproteus tenax* (Messner et al. 1986, Wildhaber and Baumeister 1987) and the surface layer of *Desulfurococcus mobilis*. *Thermoproteus tenax* was first described by Zillig et al. (1981). It is a rod-shaped bacterium of variable length and constant cell diameter (Fig.1). A regularly arrayed protein layer covers the entire cell surface, and seems to have shape-maintaining, or even shape-determining, function for this bacterium (Messner et al. 1986, Wildhaber and Baumeister 1987). This layer has p6 symmetry and an exceptionally large centre-to-centre spacing of 32.8 nm (Fig.2). It appears as a filamentous network with most of the mass accumulated around the threefold and the sixfold axes. The outer surface of the layer appears rather smooth with crater-shaped openings at the sixfold axis, whereas on the inner side a prominent protrusion, centred on the sixfold axis, extends towards the cell membrane (Figs. 1,2b), probably penetrating it completely (Wildhaber et al. 1988 in prep.). These protrusions act as spacer elements maintaining a 25-nm-wide interspace between the cell membrane and the surface protein layer itself. The existence of a distinct interspace between cell membrane and surface protein seems to be rather common amongst the archaebacteria (Wildhaber et al. 1988 in prep.). This space of unknown content may be regarded as analogous to the periplasmatic space in Gram-negative eubacteria and the surface protein may act as a molecular sieve retaining molecules beyond a certain size. However, *Desulfurococcus mobilis,* a close relative to *Thermoproteus tenax,* has a surface protein layer with p4 symmetry exhibiting very large openings (12 by 12 nm) (Wildhaber et al. 1987).

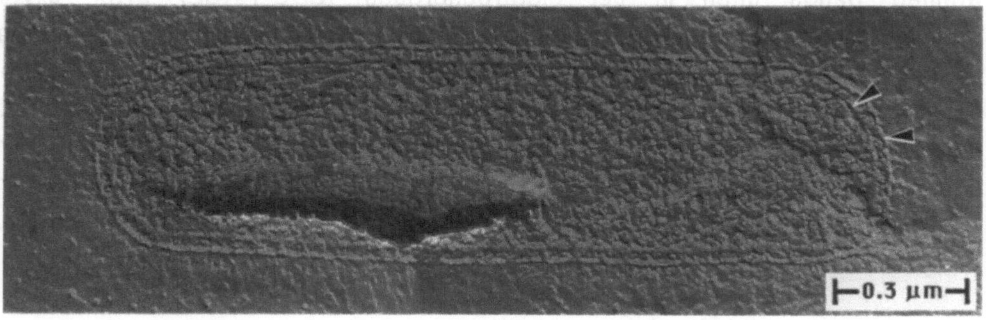

Fig.1. Freeze-etched cell of *Thermoproteus tenax* showing the typical cell wall organization: a surface protein as outermost layer with protrusions extending towards the cell membrane.
s = surface protein layer; m = cell membrane

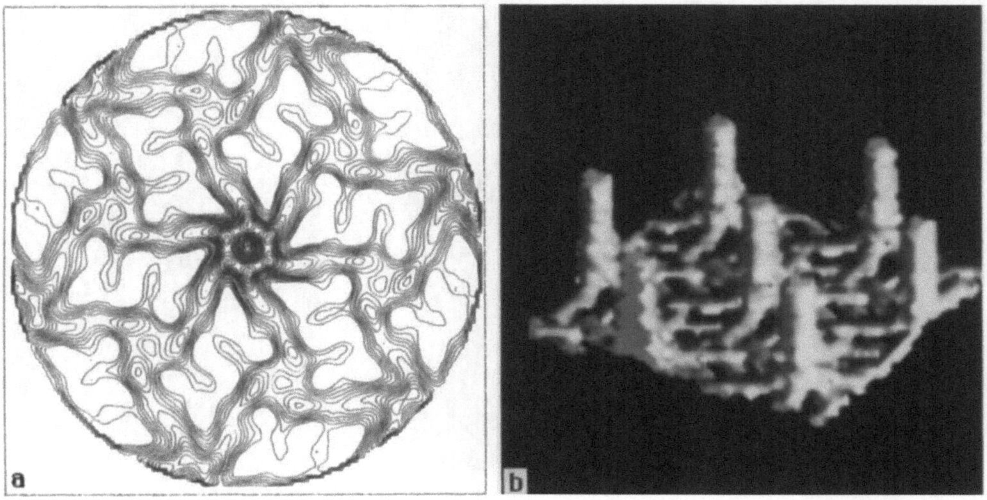

Fig.2. Average of an untilted projection from an isolated surface protein layer (a) and three-dimensional model of the layer (b). Most of the protein mass is located at the sixfold and the threefold axes

This layer can hardly be considered as an efficient sieve. Regarding the three-dimensional structure of the *Thermoproteus tenax* surface protein layer the monomer must have an extremely elongated shape, approximately 50 nm long, which corresponds to the distance from the distal end of the protrusion to the threefold axis.

All attempts to dissociate the surface protein of *Thermoproteus tenax* into the protomers have been unsuccessful so far. The faintly stained bands seen in gel electrophoreses (SDS-PAGE) after silver staining must be regarded as contaminants from other cell envelope proteins. Even extremely harsh conditions failed to dissociate it. Figure 3 shows some examples of dissociation experiments. Heating of isolated layers in presence of 2% sodium dodecyl-sulphate (SDS) for 30 min (Fig.3a), heating in 25% glycerol up to 120°C (Fig.3b) or incubation in 1 M hydrochloric acid (HCl) (Fig.3c) did not effect complete dissociation.

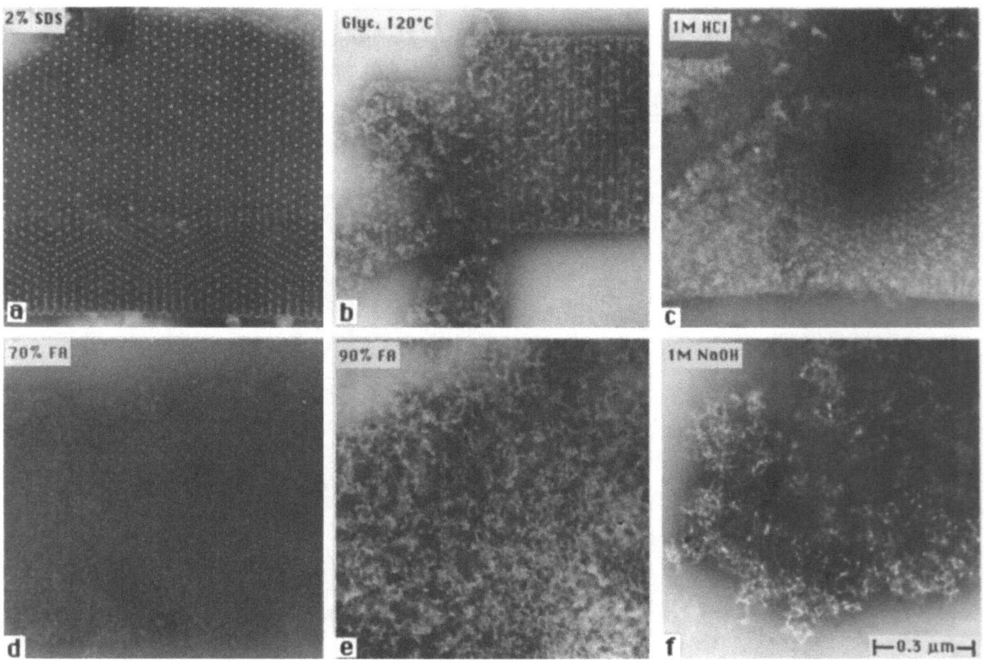

Fig.3. Isolated surface protein layers after incubation in: (a) 2% SDS at 100°C, (b) 25% glycerol at 120°C, (c) 1M HCl, (d) 70% formic acid, (e) 90% formic acid and (f) 1 M NaOH. With none of these treatments was it possible to solubilize the protein; even HCl, or heating in SDS or glycerol, did not disrupt the layer lattice completely (a-c)

Incubation in formic acid (Fig.3d 70% and Fig.3e 90%) or 1 M sodium hydroxide (NaOH) (Fig.3f) disrupted the layer lattice, but the proteins were not rendered soluble. These results could be taken as an indication that the protomeric subunits of the surface protein from *Thermoproteus tenax* are covalently cross-linked.

In order to determine the molecular weight of the monomeric subunit, STEM mass measurements (Engel et al. 1982) were carried out on isolated surface protein layers (Fig.4). After subtracting the background and calibration with tobacco mosaic virus the mass analysis resulted in 7.58 kDa/nm^2 for double layers and 3.65 kDa/nm^2 for single layers.

For crystallographic reasons a unit cell of a p6 lattice must contain six monomers (or any multiple of six). The densities described above correspond therefore to 560,000 Da/monomer and 539,500 Da/monomer respectively. Assuming a protein density of 1.3 g/cm^3 and six monomers per unit cell the molecular weight estimated from the three-dimensional model of the structure was 325,000 Da/monomer (Wildhaber and Baumeister 1987). However, models used for displaying three-dimensional structures usually underestimate the molecular volume; typically representing only 50-80% of the true volume (Baumeister and Engelhardt 1987). Moreover, contributions due to a glycosylation of the surface protein, which are taken into account in STEM mass measurements, will probably be ignored in all procedures including an averaging step because of the flexible nature of the sugar residues (Table 1).

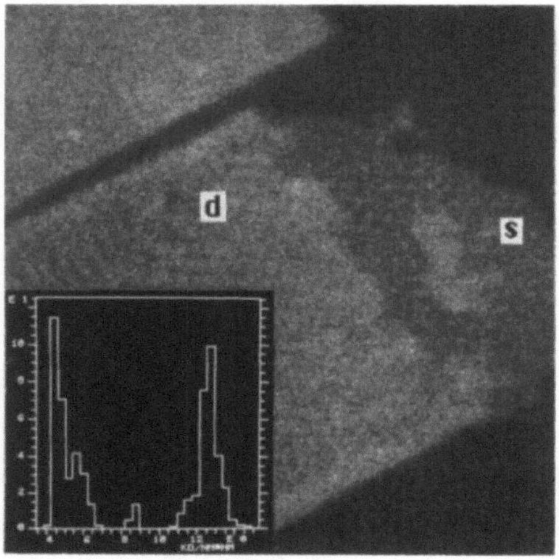

Fig.4. STEM micrograph of an isolated, unstained surface protein from *Thermoproteus tenax*; d = double layer; s = single layer. Inset: mass measurements of carbon foil (left), single layers (middle) and double layers (right). The ordinate is mass per area (kDa/nm^2) and the abscissa is the number of measurements.

Table 1. Mass determinations of the monomeric unit of the surface protein

	Mass/area (kDa/nm^2)	Volume (nm^3)	Mol wt/monomer (Da)
Background	5.03 ± 0.63 (n=305)		
S-Layer(double)	12.61 ± 0.56 (n=291)		560,000
S-Layer(single)	8.68 ± 0.15 (n=19)		539,500
3D-Structure		2.5×10^3	325,000

References

Baumeister W, Engelhardt H (1987) Three-dimensional structure of bacterial surface layers. In: Harris R, Horne RW (eds) Electron microscopy of proteins, vol 6. Membranous structures. Academic Press, London New York, pp 109-154

Engel A, Baumeister W, Saxton WO (1982) Mass mapping of a protein complex with the scanning transmission electron microscope. Proc Natl Acad Sci USA 79:4050-4054

Messner P, Pum D, Sára M, Stetter KO, Sleytr UB (1986) Ultrastructure of the cell envelope of the archaebacteria *Thermoproteus tenax* and *Thermoproteus neutrophilus.* J Bacteriol 166:1046-1054

Wildhaber I, Baumeister W (1987) The cell envelope of *Thermoproteus tenax.* Three-dimensional structure of the surface layer and its role in shape maintenance. EMBO J 6:1475-1480

Wildhaber I, Santarius U, Baumeister W (1987) Three-dimensional structure of the surface protein of *Desulfurococcus mobilis.* J Bacteriol 169:5563-5568

Wildhaber I, Studer D, Baumeister W (1988) Do surface proteins create a "periplasmic space" in archaebacteria? (in prep)

Zillig W, Stetter KO, Schäfer W, Janekovic D, Wunderl S, Holz I, Palm P (1981) Thermoproteales: a novel type of extremely thermoacidophilic anaerobic archaebacteria isolated from Icelandic solfataras. Zentralbl Bakteriol Hyg I Abt Orig C2:205-227

The Structure of Surface Proteins in *Sulfolobus*

W. Baumeister and R. Prüschenk

Max-Planck-Institut für Biochemie
D-8033 Martinsried bei München
FRG

The analysis of the three-dimensional structure of surface proteins (S-layers) from various eu- and archaebacteria has revealed a high degree of variability on their outer surface, while their inner surface appears rather more conservative (Baumeister and Engelhardt 1987). To further investigate this phenomenon we have selected several species and isolates of the genus *Sulfolobus* whose phylogenetic status is well established; the technique we have chosen for a first survey was freeze-etching, more specifically deep-etching, in conjunction with surface relief reconstruction (Prüschenk et al. 1987). This technique has the merit to provide, albeit at relatively low resolution (2.5 to 3 nm), three-dimensional information about structures in situ (Baumeister et al. 1986).

Figure 1 shows in a network representation the phylogenetic distances of four isolates *Sulfolobus solfataricus* (strain P1), *Sulfolobus* sp. (B12), *Sulfolobus acidocaldarius* (DSM 639) and *Sulfolobus* sp. (B6/2) as derived from rRNA hybridization homologies (Klenk et al. 1986). Figure 2 shows the corresponding surface relief reconstructions in a surface shading representation. Evidently, the surface reliefs fall into two classes. The first one, comprising P1 and B12 shows prominent volcano-shaped structures centred on the sixfold crystallographic axis and, at a lower level, somewhat twisted ridges along the twofold axes interconnecting the "volcanoes". The marginal differences between the surface reliefs of P1 and B12 are probably insignificant, i.e. within the limits of experimental error. The surface reliefs of B 6/2 and DSM 639 are relatively similar with respect to each other but clearly not identical; at present we cannot judge whether the minor differences emerging from the reconstructions reflect genuine differences or not. Both structures are less evident than the volcano motif in P1 or B12. The elements of the ring around the central depression are no longer contiguous; the ring, which appears much wider, is divided into six

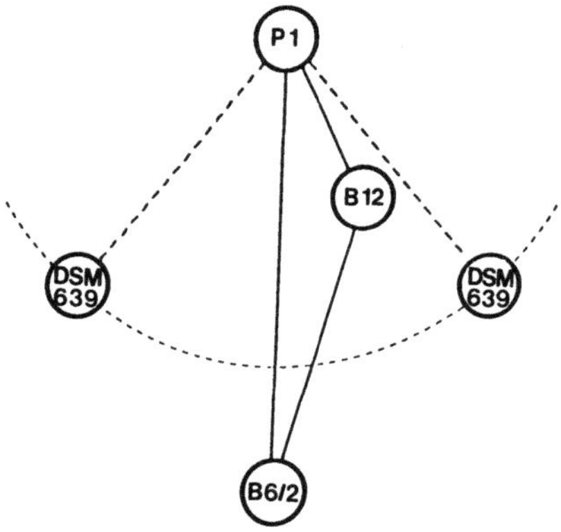

Fig.1. Network representation of phylogenetic distances of four species or isolates of *Sulfolobus* (Data from Klenk et al. 1986)

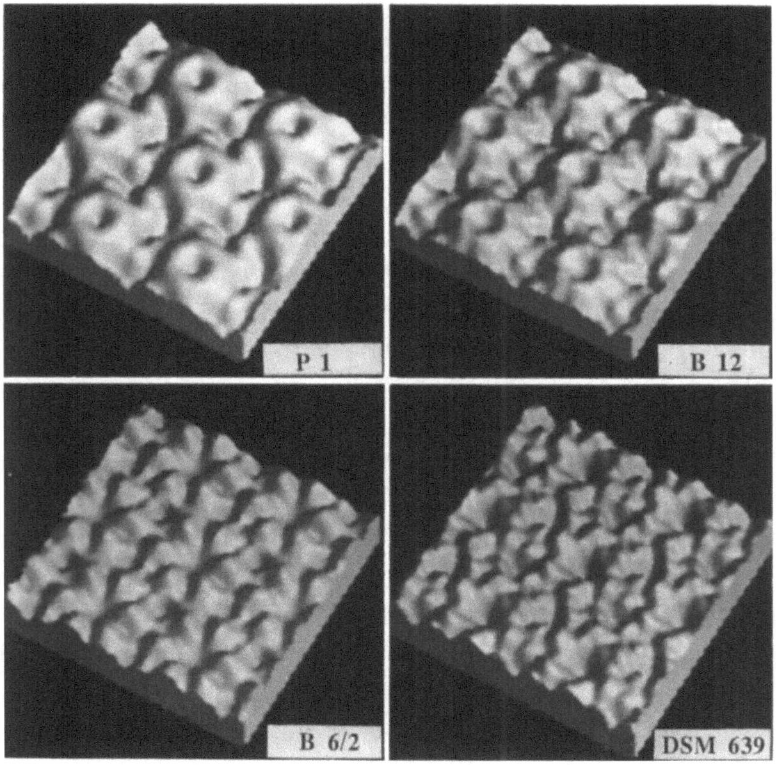

Fig.2. "Views" of surface relief reconstructions derived from deep-etching experiments with the four species or isolates of *Sulfolobus*

distinct masses. The pattern of connectivity appears to be basically the same as in P1 and in B12.

The grouping as it emerges from the surface relief reconstructions clearly reflects the phylogenetic relationships depicted in Fig.1. P1 and B12 are closely related, B6/2 is most distant relative to P1 or B12 but closer to DSM 639.

A full three-dimensional reconstruction from tilt series of negatively stained samples has been performed with the *Sulfolobus solfataricus* (P1) surface protein (Prüschenk and Baumeister 1988). Projections in z-direction through the three-dimensional reconstruction show clearly the dimeric organization of the *Sulfolobus solfataricus* layer. The "view" of the reconstructed outer surface (Fig.3) shows in greater detail now, the volcano motif already visualized by freeze-etching. The orifice on the sixfold axis opens into a "dome" which is wide open towards the inside.

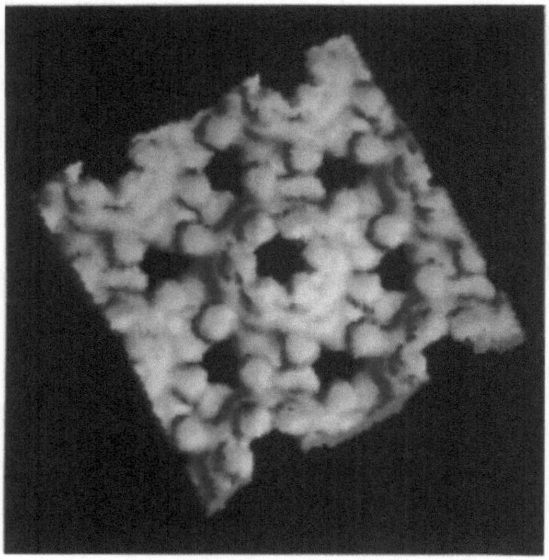

Fig.3. "View of the three-dimensional reconstruction of the *Sulfolobus solfataricus* (P1) surface layer showing its outer surface.

Since the structure of the *Sulfolobus acidocaldarius* surface protein has already been determined to a comparable resolution (Taylor et al. 1982, Deatherage et al. 1983) the two structures can be compared now section by section. The *Sulfolobus acidocaldarius* surface layer possesses on the inside a domain which is centered on the threefold axis (the T domain in Deatherage et al. 1983); this domain is completely missing in *Sulfolobus solfataricus,* a finding corroborated also by surface relief reconstructions of tantalum/tungsten shadowed S-layer sheets. Further towards the centre, the two structures are very similar to the extent even of local congruence. At a distance of approximately 2 nm from the outer surface, the two structures begin to diverge again. In *Sulfolobus acidocaldarius* strong linker elements interconnecting the hexameric units via the twofold axis reach out to the surface while in *Sulfolobus solfataricus* only the volcanoes composed of the six subunits protrude. Thus, the difference between the surface proteins of the two species is expressed on the level of their domain structure.

References

Baumeister W, Engelhardt H (1987) Three-dimensional structure of bacterial surface layers. In: Harris R, Horne RW (eds) Electron microscopy of proteins, vol 6. Membranous structures. Academic Press, London New York, pp 109-154

Baumeister W, Guckenberger R, Engelhardt H, Woodcock CLF (1986) Metal shadowing and decoration in electron microscopy of biological macromolecules. Ann N Y Acad Sci 483:57-76

Deatherage, JF, Taylor KA, Amos LA (1983) Three-dimensional arrangement of the cell wall protein of *Sulfolobus acidocaldarius.* J Mol Biol 167:823-852

Klenk HP, Haas B, Schwass V, Zillig W (1986) Hybridization homology: a new parameter for the analysis of phylogenetic relations, demonstrated with the urkingdom of the Archaebacteria. J Mol Evol 24:167-173

Prüschenk R, Baumeister W (1988) Three-dimensional structure of the surface protein of *Sulfolobus solfataricus.* Eur J Cell Biol (in press)

Prüschenk R, Baumeister W, Zillig W (1987) Surface structure variants in different species of *Sulfolobus.* FEMS Microbiol Lett 43:327-330

Taylor KA, Deatherage JF, Amos LA (1982) Structure of the S-layer of *Sulfolobus acidocaldarius.* Nature (London) 299:840-842

Structure of the Surface Glycoprotein from *Halobacterium volcanii* as Revealed by Electron Microscopy

M. Kessel[1], I. Wildhaber[2], S. Cohen[1] and W. Baumeister[2]

[1]Dept. of Membrane and Ultrastructure Research
The Hebrew University, Hadassah Medical School
P.O.Box 1172
Jerusalem 91-010, ISRAEL
[2]Max-Planck-Institut für Biochemie
D-8033 Martinsried bei München
FRG

The cell surfaces of *Halobacteria* are seen by shadow casting to be composed of a hexagonal lattice of morphological units with a centre-to-centre spacing of approximately 15 nm. Preparations of virtually salt-free envelopes from *Halobacterium volcanii,* isolated from the Dead Sea, can be obtained by freezing and thawing the cells, treatment with 1 µg/ml of DNAase and suspension in 10 mM $CaCl_2$ and have enabled a three-dimensional analysis of the surface layer by negative staining.

Envelopes of *H. volcanii* negatively stained either with 1% uranyl acetate, or 0.75% uranyl formate, both at pH 4.5, have consistently resulted in variable staining patterns irrespective of the stain used (Fig.1). One pattern is that of a "honeycomb" appearance of strong stain accumulating centres surrounded by regions of stain exclusion. Another pattern is that of a more striking negative staining with stain-excluding rings (morphological units) and their stain-accumulating centres lying in pools of stain. Optical diffractograms of these different stain patterns show hexagonal symmetry with identical lattice constants, but only patches deeply embedded in negative stain show moderately high resolution features. The two patterns may also be present in different regions of the same envelope (Fig.1) with evidence of a transition region lying between the differently stained areas. Close observation reveals that the centres of stain accumulation in the "honeycomb" arrangement lie along a sixfold axis formed by the poorly contrasted morphological units.

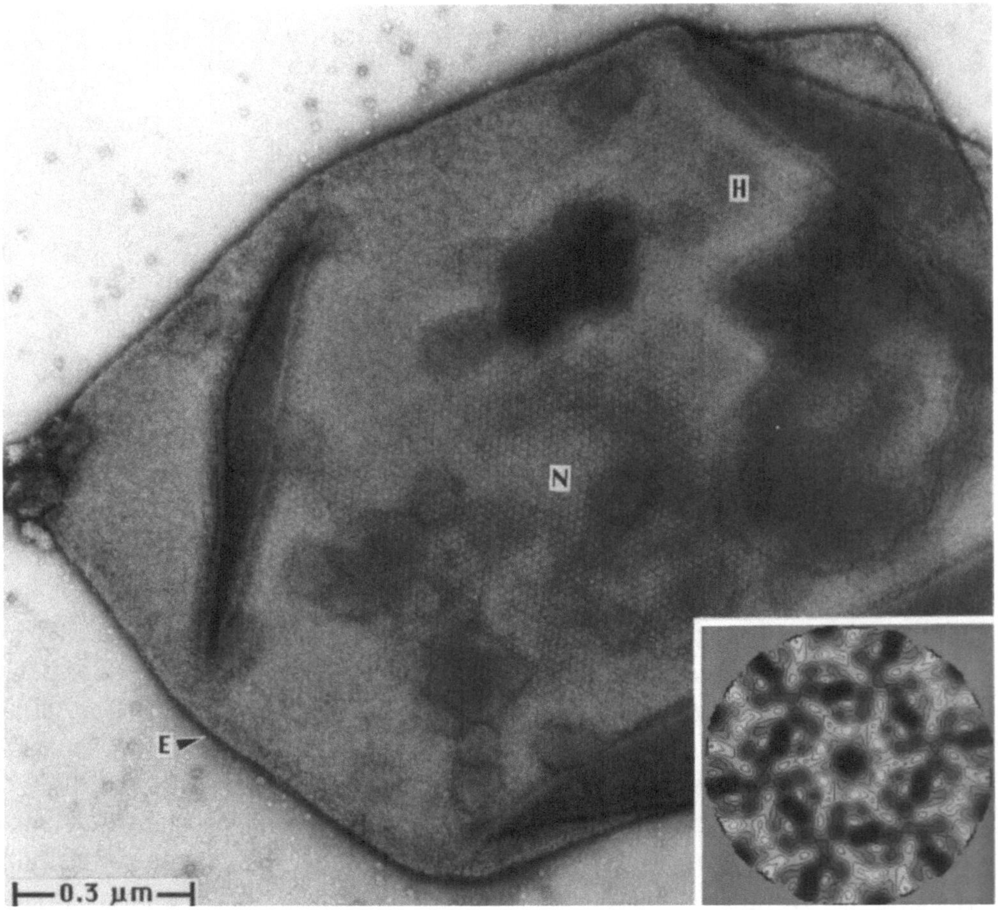

Fig.1. A typical cell envelope of *H. volcanii* negatively stained with 0.75% uranyl formate. Note the different staining patterns: (H) "honeycomb" staining; (N) well-embedded staining; (E) edge
Inset: Sixfold symmetrized correlation average from an area of the envelope well embedded in stain. Note the sixfold radial arms almost touching at the threefold axis

Edge views of morphological units when viewed at a glancing angle are seen to be located at the ends of rows of morphological units. Occasionally localized regions at the edges of the envelopes which display small arrays of well-stained morphological units, also viewed at a glancing angle along the lattice lines, are seen to fade into a transition area where the strongly staining centres remain visible surrounded by a background of stain exclusion.

Using good quality negative stain preparations we have collected a data set for three-dimensional reconstruction from a specimen tilted to 76^{0}. This high tilt angle was achieved with a special holder designed by Chalcroft and Davey (1984). Correlation averages were made from each of the 14 tilts and the data merged into a three-dimensional data set by the hybrid Fourier space/real space method introduced by Saxton et al. (1984).

The three-dimensional reconstruction shows the morphological unit to be a dome shaped structure, 9 nm in diameter, composed of six subunits with a small orifice at the apex of a small depression in the dome (Fig.2). The orifice opens into a funnel-shaped pore which spans the 4.5 nm height of the dome. Six radial protrusions emanate from each morphological unit and appear to meet at the threefold symmetry axis at a height of 2.7 nm and provide the lateral connections holding the surface layer intact. The resolution of the reconstruction is of the order of 2 nm and at this resolution these protrusions are not resolved as touching. Additional smaller protrusions, pointing downwards, are located between the radial arms. Freeze fracture of the cell surface confirms that the morphological units are dome shaped and in favourable orientations to the direction of shadow small depressions are seen at the centre of each morphological unit. The dome, therefore, faces outwards and is the structural element of the cell facing the surrounding milieu. Edge views show the wall morphological units as adjacent, 4.5 nm high, parabolas confirming that the dome points outwards. Based on this polarity the inverted funnel terminates in the region of the cell

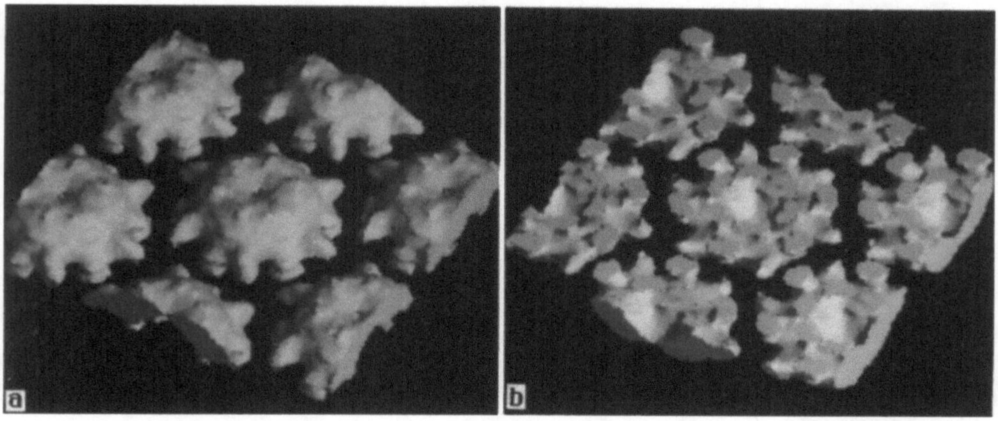

Fig.2. (a) The three-dimensional model viewed from the top (outside of the cell) with a vertical section emphasizing the dome shape of the morphological unit and the funnel-like pore. (b) The three-dimensional model viewed from the "inside"

membrane. Our reconstruction apparently suffers from a lack of definition in the region of the membrane, most likely due to a flattening effect of the negative staining. The combined results of: (1) our observations on the staining behaviour of the *H. volcanii* envelopes and the resulting three-dimensional reconstruction; (2) the evidence of the organization of the layers in the envelope from *H. halobium* by X-ray diffraction (Blaurock et al. 1976); and most recently, (3) the cloning of the gene of the cell surface glycoprotein in *H. halobium* and thereby the knowledge of the amino acid sequence (Lechner and Sumper 1987) and exact sites of glycosylation (Paul and Wieland 1987) in a closely related organism, have been integrated into a model for the organization of the *Halobacterium* envelope (Kessel et al. 1988 submitted).

References

Blaurock A, Stoeckenius W, Oesterhelt D, Scherphof G (1976) Structure of the cell envelope of *Halobacterium halobium.* J Cell Biol 71:1-22

Chalcroft J, Davey CL (1984) A simply constructed extreme-tilt holder for the Philips eucentric goniometer stage. J Microsc 134:41-48

Kessel M, Wildhaber I, Cohen S, Baumeister W (1988) Three dimensional structure of the regular surface glycoprotein layer of *Halobacterium volcanii* from the Dead Sea. Proc Natl Acad Sci USA (submitted)

Lechner J, Sumper M (1987) The primary structure of a procaryotic glycoprotein. J Biol Chem 262: 9724-9729

Paul G, Wieland F (1987) Sequence of the halobacterial glycosaminoglycan. J Biol Chem 262:9587-9593

Saxton WO, Baumeister W, Hahn M (1984) Three-dimensional reconstruction of imperfect two-dimensional crystals. Ultramicroscopy 13:57-70

The Structure of the Surface Protein of *Thermotoga maritima*

R. Rachel[1], I. Wildhaber[1], K.O. Stetter[2] and W. Baumeister[1]

[1]Max-Planck-Institut für Biochemie
D-8033 Martinsried bei München, FRG
[2]Lehrstuhl für Mikrobiologie, Universität Regensburg
D-8400 Regensburg, FRG

Thermotoga maritima has recently been described as an extremely thermophilic eubacterium growing at temperatures as high as 90°C (Huber et al. 1986). According to 16s rRNA sequencing it represents the deepest known branch and the most slowly evolving line within the kingdom of eubacteria (Achenbach-Richter et al. 1987) (Fig.1). Because also most members of the second deepest branch within the eubacteria, the green non-sulphur bacteria, including *Thermomicrobium, Chloroflexus* and *Herpetosiphon* (Oyaizu et al. 1987), are thermophilic, the hypothesis has been put forward that the "original eubacteria" were thermophiles (Achenbach-Richter et al. 1987). The cells of *Thermotoga maritima* are loosely surrounded by a sheath-like envelope which tends to balloon at the cell poles. In this communication we present some results of an as yet rather preliminary analysis of the structure of this outer sheath.

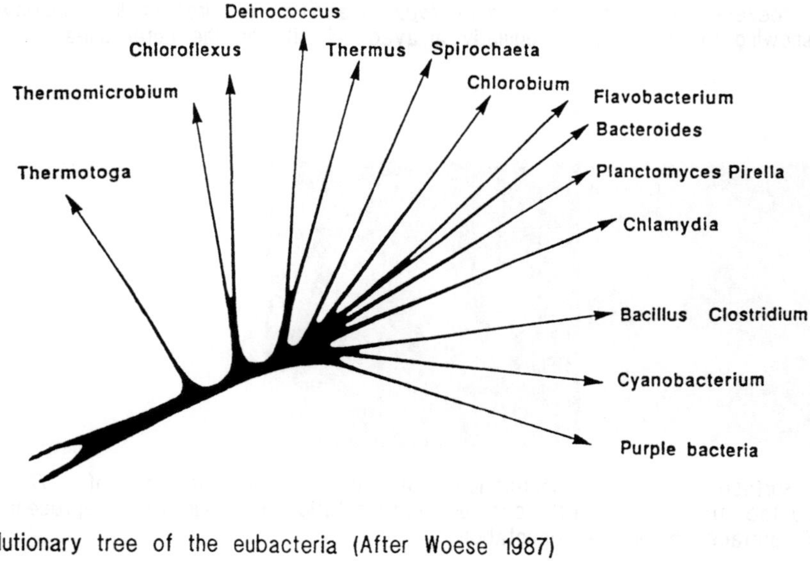

Fig.1. Evolutionary tree of the eubacteria (After Woese 1987)

On electron micrographs of freeze-fractured (Huber et al. 1986) or freeze-etched (Fig.2) cells (unidirectionally shadowed with Pt/C) the outer sheath is visible as being composed of regularly arrayed subunits. Visual inspection of the micrographs and optical diffraction analysis clearly show that the size of the coherent crystalline patches in the sheaths is rather small, thus imposing some problems for a high resolution analysis of this specimen. The micrographs have been analyzed by surface relief reconstruction (Guckenberger 1985) after correlation averaging over small patches. The reconstruction (Fig.3) reveals three prominent depressions centred around a threefold symmetry axis. Elevations on two other threefold axes surround these depressions. This motif, which strikingly resembles that found in the analysis of electron micrographs of reconstituted *E. coli* porins (Chang et al. 1985, Dorset

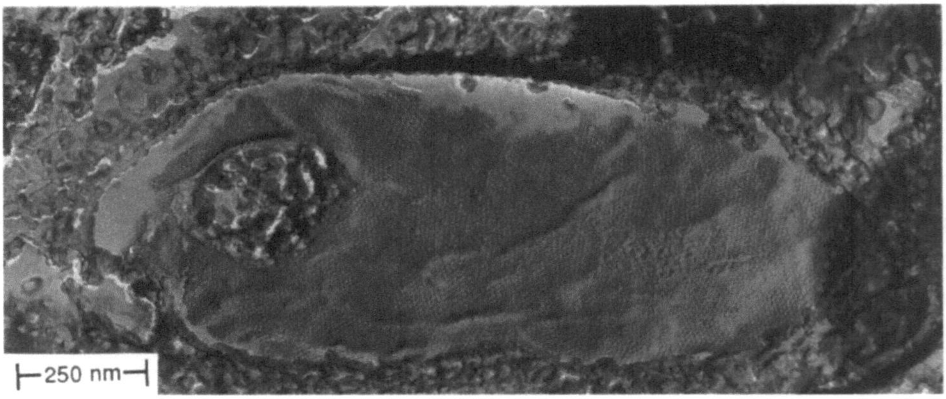

Fig.2. Freeze-etched cell of *Thermotoga maritima* (unidirectionally shadowed with Pt/C) showing the patches of regularly arrayed subunits on the outer sheath

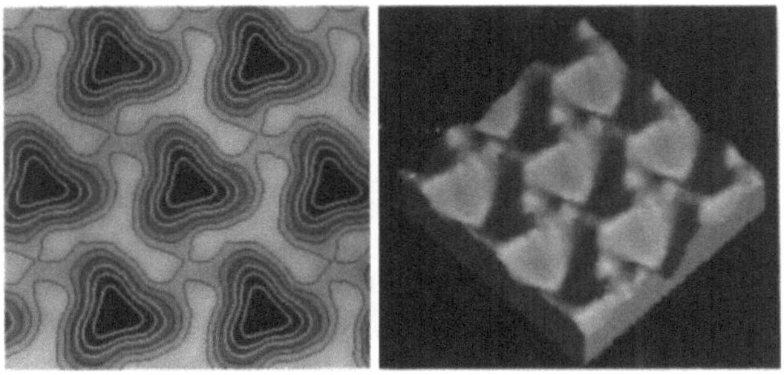

Fig.3. Surface relief reconstruction of the outer sheath of *Th. maritima*, reconstructed from Fig.2. Left: contour representation (contour lines represent heights); right: 3D surface shading representation

et al. 1983) and natural assemblies of porin-like molecules (Kessel et al. 1985, Chalcroft et al. 1987), constitutes a p3 lattice with a centre-to-centre spacing of approximately 11 nm. For a further structural and chemical characterization, we attempted to isolate and purify the outer sheath of *Thermotoga maritima.* Cells harvested in the exponential phase of growth were disrupted by sonication (35 W; 5 min) in a buffer containing triethanolamine-HCl, pH 5.8 (25 mM), $MgCl_2$ (2 mM), PMSF (0.2 mM), and NaN_3 (2 mM). The envelopes were collected by centrifugation (30,000 x *g*) and subsequently fractionated on a Percoll density gradient (30% v/v). A thin but distinct band highly enriched in the outer sheath was found at a buoyant density of approx. 1.020 g/cm^3. The main protein component of this fraction had an apparent molecular weight of 43,000 Da as determined on SDS-polyacrylamide gel electrophoresis. The analysis of this material by electron microscopy after negative staining with Na-phosphotungstate confirmed its homogeneity; it contained almost exclusively fragments of the outer sheath (Fig.4a).

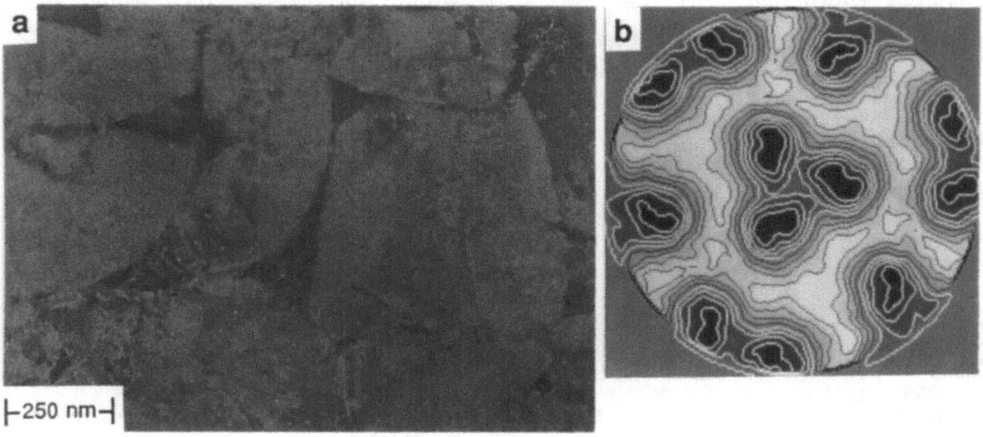

Fig.4. (a) Preparation of isolated outer sheaths of *Th. maritima* (negatively stained). (b) Correlation average of the outer sheath including threefold symmetrization

Small areas of the micrograph were digitized and processed by correlation averaging (Saxton and Baumeister 1982). The averages (Fig.4b) confirmed the result of the surface relief reconstruction (Fig.3), showing the sheath to be composed of subunits with p3 symmetry. The averaged unit cell derived from micrographs of negatively stained outer sheaths (Fig.4b) may tentatively be interpreted as comprising three stain-filled pores, surrounded by a network of stain-excluding material, most likely protein and lipid. Again the similarity between the subunit structure of the outer sheath of *Th. maritima* and that of several outer membrane proteins is quite obvious. Also the apparent M_r of the main protein component (43,000 Da) is in the range found for OMP's in many other bacteria (Hancock 1987).

All the results obtained so far with the outer sheath of *Th. maritima* - the buoyant density, subunit structure in electron micrographs, and apparent M_r of the assumed monomer - suggest categoring it as a rOMP (Baumeister and Engelhardt 1987) rather than as a bona fide S-layer. However, this is a rather preliminary interpretation and more conclusive evidence has to be accumulated. For a more rigorous comparison, a 3D reconstruction obtained by a tilt series in the electron microscopy, a determination of the lipid-to-protein ratio in the outer sheath, a determination of the secondary structure of the 43,000 Da protein, and possibly sequence data are required. If these further investigations on the outer sheath of *Th. maritima* will indeed support a close relationship to the OMP's of other eubacteria, this would have interesting implications for the evolution of cell envelopes in eubacteria.

References

Achenbach-Richter L, Gupta R, Stetter KO, Woese CR (1987) Were the original eubacteria thermophiles? Syst Appl Microbiol 9:34-39

Baumeister W, Engelhardt H (1987) Three-dimensional structure of bacterial surface layers. In: Harris JR, Horne RW (eds) Electron microscopy of proteins, vol 6. Membranous structures. Academic Press, London New York, pp 109-154

Chalcroft JP, Engelhardt H, Baumeister W (1987) Structure of the porin from a bacterial stalk. FEBS Lett 211:53-58

Chang CF, Mizushima S, Glaeser RM (1985) Projected structure of the pore-forming OMPC protein from *Escherichia coli* outer membrane. Biophys J 47:629-639

Dorset DL, Engel A, Haner M, Massalski A, Rosenbusch JP (1983) Two-dimensional crystal packing of matrix porin. J Mol Biol 165:701-710

Guckenberger R (1985) Surface reliefs derived from heavy-metal-shadowed specimens. Fourier space techniques applied to periodic objects. Ultramicroscopy 16:357-370

Hancock REW (1987) Model membrane studies of porin function. In: Inouye M (ed) Bacterial outer membranes as model systems. Wiley, New York, pp 187-225

Huber R, Langworthy TA, König H, Thomm M, Woese CR, Sleytr UB, Stetter KO (1986) *Thermotoga maritima* sp. nov. represents a new genus of unique extremely thermophilic eubacteria growing up to 90°C. Arch Microbiol 144:324-333

Kessel M, Radermacher M, Frank J (1985) The structure of the stalk surface layer of a brine pond microorganism: correlation averaging applied to a double layered lattice structure. J Microsc 139:63-74

Oyaizu H, Debrunner-Vossbrinck B, Mandelco L, Studier JA, Woese CR (1987) The green non-sulfur bacteria: a deep branching in the eubacterial line of descent. Syst Appl Microbiol 9:47-53

Saxton WO, Baumeister W (1982) The correlation averaging of a regularly arranged bacterial cell envelope protein. J Microsc 127:127-138

Woese CR (1987) Bacterial evolution. Microbiol Rev 51:221-271

Three-Dimensional Structure of Surface Layers from Various *Bacillus* and *Clostridium* Species

H. Engelhardt, Z. Cejka and W. Baumeister

Max-Planck-Institut für Biochemie
D-8033 Martinsried bei München
FRG

The analysis of the projected structure of S-layers has become a standard technique in recent years. It revealed basic structural features such as symmetry, apparent porosity and mass disposition in the projected unit cell and allowed to propose a classification of S-layers which is of some heuristic value (Baumeister et al. 1986). However, 3-D reconstructions are required (besides primary structure information) to describe and to compare the S-layers in greater detail. Here, the 3-D structures of S-layers of various *Bacillus* and *Clostridium* species are compared that are phylogenetically distant on the species, the genus, and the family level (Fig.1).

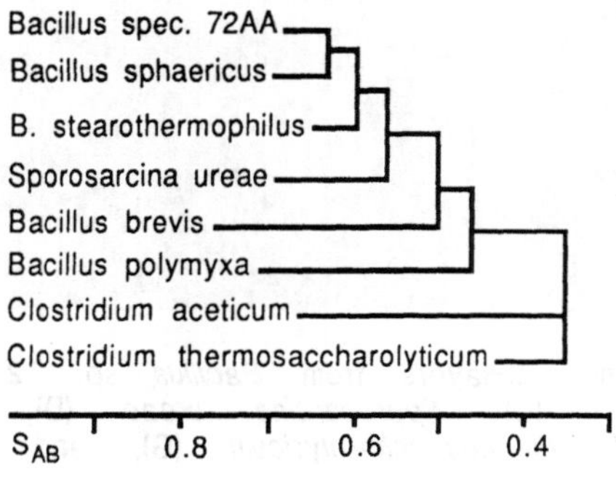

Fig.1. Phylogenetic relationships among various *Bacillus* and *Clostridium* species deduced from 16s rRNA oligonucleotide cataloguing. (Courtesy of E. Stackebrandt, Kiel)

The S-layers of several bacilli, including *Sporosarcina ureae,* are almost indistinguishable in projection (Fig.2, A-D) except for *B. sphaericus* P1 (Lepault et al. 1986) whose S-layer exhibits a slightly different mass distribution but the same basic architecture. The S-layer proteins, however, are not identical. The behaviour in lithium dodecylsulphate PAGE, as well as the band patterns in gels obtained by peptide mapping (Fig.3), reflect differences on the primary structure

The peptide map of the *B. polymyxa* layer protein, on the other hand, is clearly different. This correlates with a different mass disposition in the unit cell and a smaller lattice constant (Fig.2). According to the similarity coefficient of 16s rRNA cataloguing (S_{AB} 0.45) *B. polymyxa* and *B. sphaericus* and relatives are already separated on the genus level (Fig.1). The 3-D reconstructions of the *Bacillus* S-layers show that the inner surfaces, i.e. the core regions, are structurally conservative (Fig.4). This observation applies to most other S-layers reconstructed so far (Baumeister and Engelhardt 1987). The domain structure on the outer surfaces, however, are variable and apparently species-specific (Figs.4,5). The 3-D reconstructions of the *B.* sp. 72AA (unpublished) and the *S. ureae* S-layers (Engelhardt et al. 1986), too, are not identical concerning the structure of the outer surface though the projections are indistinguishable (Fig.2).

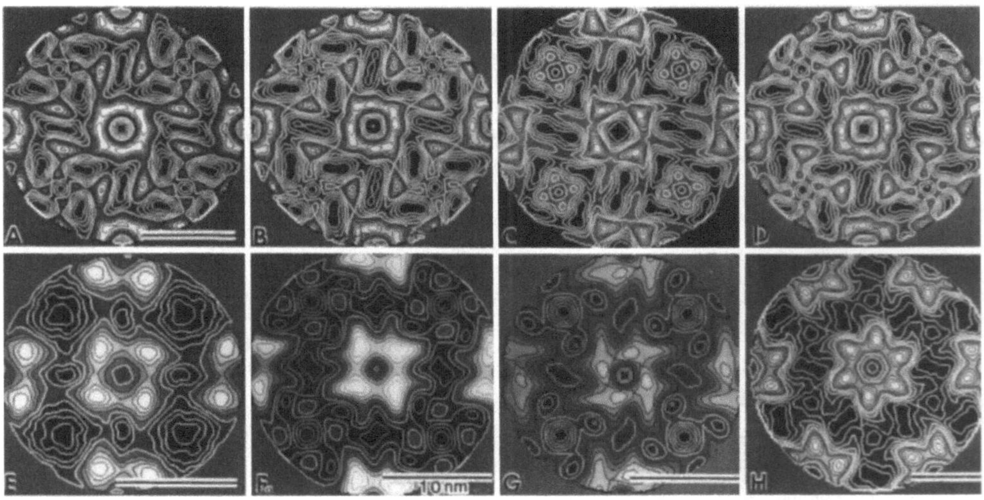

Fig.2. Correlation averages of negatively stained S-layers from *Bacillus* sp. 72AA (a), *B. sphaericus* (B), *B. sphaericus* P-1 (C), *Sporosarcina ureae* (D), *B. polymyxa* (E), *Clostridium aceticum* (F), *C. thermosaccharolyticum* (G), and *C. thermohydrosulfuricum* (H). Bars = 10 nm

The tetragonal layers of *Clostridium aceticum* (Woodcock et al. 1986) and *C. thermosaccharolyticum* (Cejka and Baumeister 1987) are very similar in projection although the species are phylogenetically distant (Fig.1). Again, the 3-D reconstructions revealed significant differences (Fig.4). The core on the inner surface of the *C. thermosaccharolyticum* layer is much more massive, while the domains providing connectivity

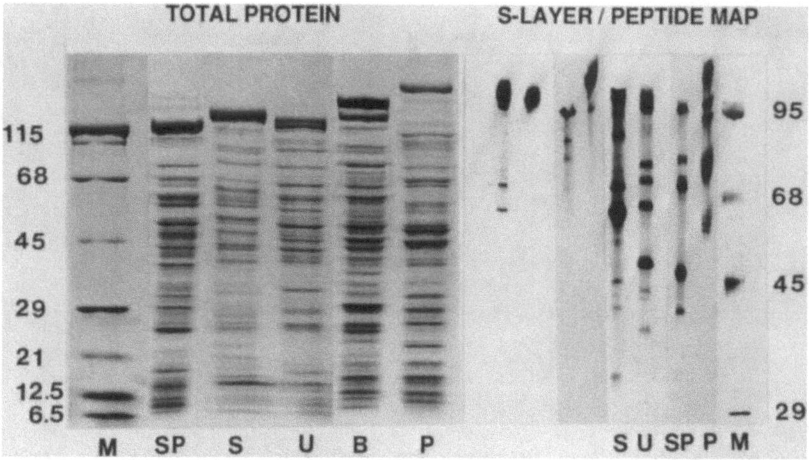

Fig.3. LDS-PAGE of the total cell protein of various *Bacillus* species. The prominent high molecular bands represent the S-layer proteins (Coomassie blue stained). Peptide mapping of the S-layer proteins using N-Chlorosuccinimide (at the right), untreated samples (at the left; silver stained). M: marker proteins, apparent M_r indicated (x 1/1000), SP: *B.* sp. 72AA, S: *B. sphaericus,* U: *S. ureae,* B: *B. brevis,* P: *B. polymyxa*

between adjacent unit cells are clearly fainter than the corresponding structures of the *C. aceticum* S-layer; the two layers are indeed very dissimilar with respect to their spatial organization.

While the tetragonal *Bacillus* and *Clostridium* S-layers may be regarded as homologous, i.e. probably deriving from a common ancestral protein, the question arises whether the S-layer proteins creating lattices with other symmetries may also be homologous. *B. stearothermophilus* (Messner et al. 1984) and *B. brevis* (Tsukagoshi et al. 1982) possess S-layers with p2- and p6-symmetry, respectively, and are closely related to the *Bacillus* species considered here (Fig.1). Comparing the 3-D reconstructions of the S-layers from *C. aceticum* and *C. thermohydrosulfuricum* (Cejka et al. 1986) it becomes obvious that the gross morphology (disregarding the symmetries) is surprisingly similar (Fig.5). This observation also applies to the (putative) monomers, if a domain in the core region of the *C. aceticum* layer protein is removed ("deletion"). Doing so, the mut(il)ated monomers can be rearranged on a hexagonal lattice such that an S-layer with a centre-to-centre spacing of the original *C. thermohydrosulfuricum* array is obtained (Woodcock et al. 1986).

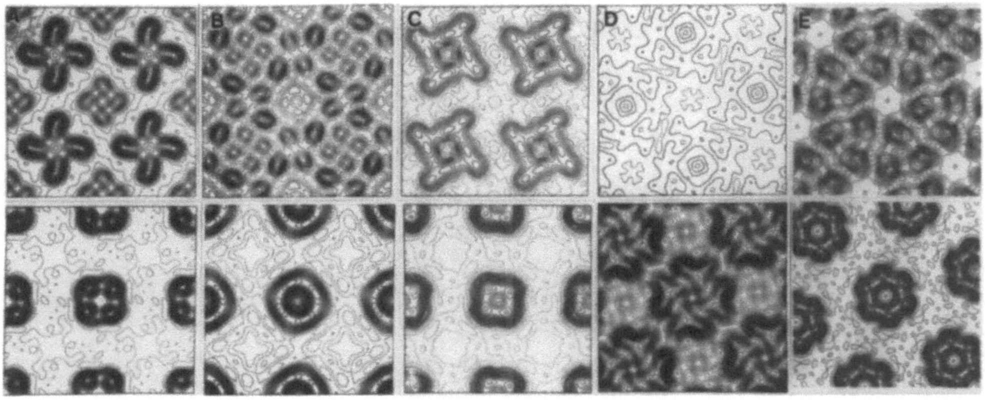

Fig.4. Horizontal sections through the 3-D reconstruction of the S-layers near the outer surface (top row) and near the inner surface (bottom row) from *S. ureae* (A), *B. sphaericus* *P1 (B), *C. aceticum* (C), *C. thermosaccharolyticum* (D), and *C. thermohydrosulfuricum* (E)

(*3-D data from Lepault et al. 1986)

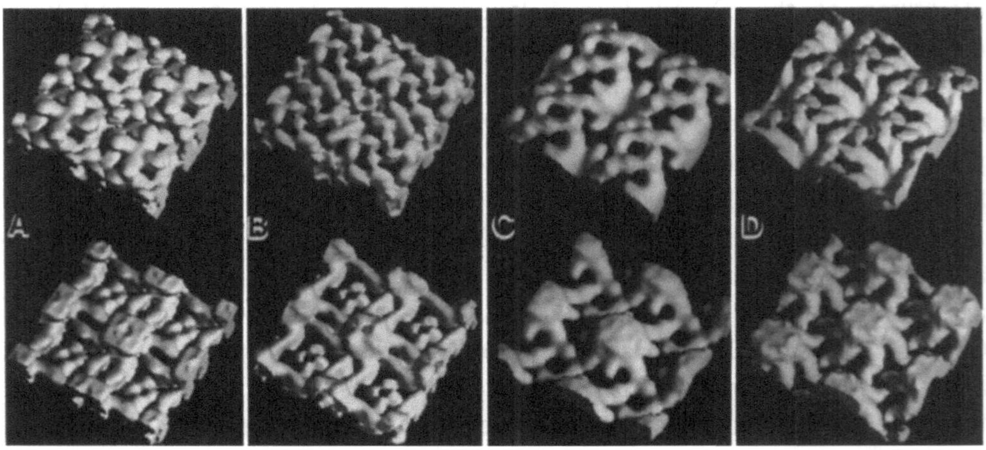

Fig.5. Three-dimensional representation of the S-layers of *S. ureae* (A), *B. sphaericus* *P-1 (B), *C. aceticum* (C), and *C. thermohydrosulfuricum* (D). Top row: outer surfaces, bottom row: inner surfaces

(*3-D data from Lepault et al. 1986)

The following conclusions can be drawn:

- The outer surface of (eu-)bacterial S-layers is species-specific on the domain structure level; the biological implications such as specific cell-cell recognition are discussed in Baumeister and Engelhardt (1987).
- The inner surface is structurally conservative, indicating the evolutionary influence of the underlying cell wall and its interaction with the S-layer.
- S-layer proteins, forming different crystals, are not implicitly unrelated.
- S-layers can only be compared reliably on the 3-D structure level. Projections may suggest non-existing similarities or conceal common features.

References

Baumeister W, Engelhardt H (1987) Three-dimensional structure of bacterial surface layers. In: Harris JR, Horne RW (eds) Electron microscopy of proteins, vol 6. Membranous structures. Academic Press, London New York, pp 109-154

Baumeister W, Barth M, Hegerl R, Guckenberger R, Hahn M, Saxton WO (1986) Three-dimensional structure of the regular surface layer (HPI-layer) of *Deinococcus radiodurans.* J Mol Biol 187:241-253

Cejka Z, Baumeister W (1987) Three-dimensional structure of the surface protein of *Clostridium thermosaccharolyticum.* FEMS Microbiol Lett 44:13-18

Cejka Z, Hegerl R, Baumeister W (1986) Three-dimensional structure of the surface layer protein of *Clostridium thermohydrosulfuricum.* J Ultrastruct Mol Struct Res 96:1-11

Engelhardt H, Saxton WO, Baumeister W (1986) Three-dimensional structure of the tetragonal surface layer of *Sporosarcina ureae.* J Bacteriol 168:309-317

Lepault J, Martin N, Leonard K (1986) Three-dimensional structure of the T-layer of *Bacillus sphaericus* P-1. J Bacteriol 168:303-308

Messner P, Hollaus F, Sleytr UB (1984) Paracrystalline cell wall surface layer of different *Bacillus stearothermophilus* strains. Int J Syst Bacteriol 34:202-210

Tsukagoshi N, Yamada H, Tsuboi A, Udaka S, Katsura I (1982) Hexagonal surface array in a protein-secreting bacterium, *Bacillus brevis* 47. Biochim Biophys Acta 693:134-142

Woodcock CLF, Engelhardt H, Baumeister W (1986) The tetragonal surface layer of *Clostridium aceticum:* three-dimensional structure and comparison with the hexagonal layer of *Clostridium thermohydrosulfuricum.* Eur J Cell Biol 42:211-217

Paracrystalline Surface Layer of a Novel *Bacillus* sp M3198

M. Haapasalo[1], K. Lounatmaa[2,3] and H. Ranta[1,4]

[1]Department of Cariology, [2]Department of General Microbiology and
[3]Department of Electron Microscopy
University of Helsinki
Mannerheimintie 172 and
[4]National Public Health Institute
SF-00280 Helsinki 28
FINLAND

Periodic structures composed of crystalline proteins (S-layers) have been described on cell surfaces of many bacteria (Sleytr and Messner 1983). Information on the occurrence of the S-layers in different species is needed for the evaluation of its possible role for bacterial ecology and in infection. In this chapter we describe the characterization of a novel *Bacillus* sp. strain M3198. For comparison, type strains of biochemically closely related species, *B. coagulans* and *B. circulans* were also studied.

Methods

Cultural conditions. Strain M3198 was isolated together with anaerobic bacteria from a dental root canal infection. Possibility for airborne contamination could not be excluded. Pure cultures were preserved in glycerine-milk at -80°C. For biochemical tests and for ultrastructural studies strains M31898, *B. coagulans* DSM1, and *B. circulans* DSM11 were grown on blood agar plate rich in nutrients (MCG-plate, Haapasalo 1986) in air at 37°C.

Biochemical tests. The strains were studied following the identification scheme given by Sneath (1986). Fermentations were done by the API 50CH system. Glycosidases and aminopeptidases were screened with Diagnostic Tablets (Rosco, Denmark) and with RapID ANA system (Innovative Diagnostic System, USA). Hydrolysis of starch, urea and esculin, liquefaction of gelatin, production of indole, nitrate reduction, and lipase and lecithinase production were also studied (Phillips and Nash 1985).

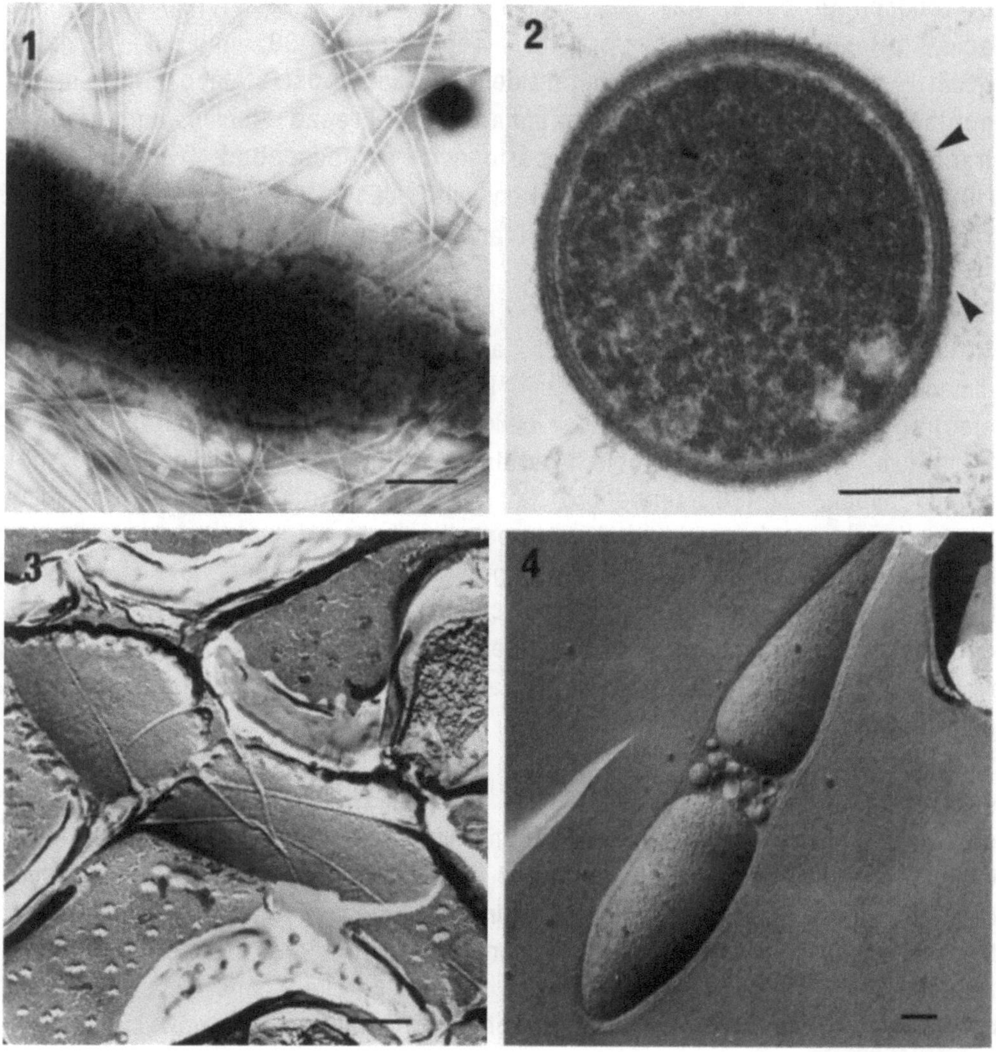

Fig.1. A negatively stained cell of *Bacillus* sp. M3198. The cell wall and flagella are seen, but no periodicity could be observed with this technique. Bar in all the micrographs = 0.2 µm

Fig.2. A thin sectioned cell of M3198. The cytoplasm is covered by the multilayered cell wall. Periodicity can be seen in some areas (arrows)

Fig.3. A freeze-etched cell with crossed flagella on the S-layer with linear (oblique)-looking periodicity

Fig.4. A freeze-fractured cell with the convex fracture face of the cytoplasmic membrane. Two rows of particles around the fractured cell represent cell wall and the S-layer

Electron microscopy. For negative staining, bacteria were suspended in 0.1 phosphate buffer (pH 7.2) and stained with 1% (w/v) phosphotungstic acid (pH 6.5). Samples for thin sections (M3198 and *B. circulans* DSM11) were pre-fixed in glutaraldehyde and post-stained with uranyl acetate and lead citrate (Lounatmaa 1985). Freeze-etching and freeze-fracture of strain M3198 was performed as described by Lounatmaa (1985). The micrographs were taken with JEM-100CX and JEM-1200EX transmission electron microscopes at an operating voltage of 60 kV.

Results and Discussion

M3198 grew in both anaerobic and aerobic atmosphere. The cell dimensions were 0.5 μm by 1-4 μm. It always stained Gram-negative, but electron microscopy revealed a cell wall structure of a Gram-positive bacterium. M3198 was motile, catalase-positive, but spores were not clearly seen. M3198 fermented a variety of carbohydrates including xylose and arabinose, hydrolyzed esculin, but did not produce lipase, lecithinase, gelatinase or indole. It grew at +45°C but not at +55°C. Biochemical characteristics were closest to *B. coagulans* and *B. circulans,* but comparison with the type strains of these species revealed differences in fermentation and enzyme patterns, and in colonial characteristics.

Electron microscope findings are shown in Figs. 1-4. The oblique symmetry of the S-layer of M3198 was clearly seen in freeze-etched cells. The distance between adjacent lines was approximately 9.5 nm. Oblique symmetry has been reported for some strains of *B. sphaericus,* but it seems to be uncommon in the genus *Bacillus* (Sleytr and Messner 1983). Neither S-layer nor periodicity was observed in *B. coagulans* or *B. circulans* strains. More detailed analysis of strain M3198 is needed to evaluate the taxonomic position of the new isolate.

References

Haapasalo M (1986) *Bacteroides buccae* and related taxa in necrotic root canal infections. J Clin Microbiol 24:940-944

Lounatmaa K (1985) Electron microscopic methods for the study of bacterial surface structures. In: Korhonen TK, Dawes EA, Makela PH (eds) Enterobacterial surface antigens: methods for molecular characterization. Elsevier, Amsterdam, pp 243-261

Phillips E, Nash P (1985) Culture media. In: Lennette EH, Balows A, Hausler WJ Jr, Shadomy HJ (eds) Manual of clinical microbiology, 4th edn. Am Soc Microbiol, Washington DC, pp 1051-1092

Sleytr UB, Messner P (1983) Crystalline surface layers on bacteria. Annu Rev Microbiol 37:311-339

Sneath PHA (1986) Endospore-forming Gram-positive rods and cocci. In: Sneath PHA, Mair NS, Sharpe ME, Holt JG (eds) Bergey's manual of determinative bacteriology, vol II, 9th edn. Williams & Wilkins, Baltimore, pp 1104-1207

Three-Dimensional Structure of the Regular Tetragonal Surface Layer of *Azotobacter vinelandii*

W.H. Bingle[3], H. Engelhardt[2], W.J. Page[1] and W. Baumeister[2]

[1]Department of Microbiology
University of Alberta
Edmonton, Alberta, T6G 2E9, CANADA
[2]Max-Planck-Institut für Biochemie
D-8033 Martinsried bei München, FRG
[3]Department of Microbiology
University of British Columbia
Vancouver, V6T 1W5, CANADA

Azotobacter vinelandii is an obligately aerobic, nitrogen-fixing soil bacterium closely related to the fluorescent pseudomonads. The subject of numerous electron microscopic studies during the past 20 years, this organism was only recently shown to possess a regular tetragonal S-layer (Bingle et al. 1984). Conventional Fourier-based image enhancement methods have revealed its projection structure (Bingle et al. 1987), which is quite similar to the tetragonal S-layer of *Aeromonas salmonicida* (Stewart et al. 1986). We report here the 3-D structure of the *A. vinelandii* S-layer using a reconstruction scheme which involves the use of "correlation averaging", an image-processing technique which unlike conventional Fourier-based methods, partially compensates for lattice distortions (Saxton and Baumeister 1982). In addition, fragments of the S-layer free of underlying outer membrane material have been obtained for the first time. These two improvements have allowed the determination of the 3-D structure of the *A. vinelandii* S-layer to a resolution of 1.7 nm.

Electron Microscopy and Image Processing Methods

Electron micrographs were made using a Philips EM 420 equipped with an unlimited tilt specimen holder. Tilt series were recorded following a scheme outlined by Saxton et al. (1984). Regions of the negatives containing S-layer fragments exhibiting satisfactory crystallinity were digitized by densitometry in a matrix of 512^2 pixels with a step size of

20 µm; this corresponds to a translation of 0.56 nm at the specimen level. Finally, all images were subjected to correlation averaging (Saxton 1985) in order to define the average unit cell. The three-dimensional reconstruction was performed as outlined in detail by Saxton et al. (1984) and Baumeister et al. (1986a). The sidedness of the S-layer was determined by performing a surface relief reconstruction of unidirectionally platinum/carbon shadowed freeze-etched whole cells. Freeze-etched cells were subjected to correlation averaging and subsequently one of them to correspondence analysis (Frank and van Heel 1982). Correspondence analysis was performed according to a variant described previously (Engelhardt et al. 1985) using 480 unit cells extracted from the source image. Surface relief reconstruction was performed as described by Guckenberger (1985) and Baumeister et al. (1986b) starting with the selectively averaged unit cells. Both the 3D-model and the surface relief reconstruction were visualized using a surface shading program which essentially produces simulated photographs of the S-layer as if it possessed a reflective surface and was illuminated from the right at an angle of 45^{0} (Saxton 1985).

Projection Structure of the *A. vinelandii* S-Layer

Preparation of *A. vinelandii* OP strain UW (capsule^{-}) "ghosts" (Robish and Marr 1962) yielded large particles retaining the shape of the cell possessing one, large, centrally located hole where osmotic disruption presumably occurred (Fig.1A). While the tetragonal S-layer was clearly visible covering the entire ghost, optical diffractograms indicated that these areas were composed of superimposed lattices and that the separation of the lattices for correlation averaging would be a formidable task. Therefore, single-layered areas of the S-layer visible through the "holes" in the cell wall were averaged. Correlation averaging of a number of such holes produced in projection, at a resolution of 2 to 2.5 nm, a pattern of two alternating interconnected tetrameric subunits (Fig.1B). The lattice constant based on three determinations was a = b = 12.5+/-0.3 nm.

Large fragments of the S-layer free of the outer membrane were obtained by exposing whole cells to 100 uM EDTA in 100 mM Tris-HCl, pH 8. Correlation averaging of this material produced a projection structure (resolution 1.7 nm) with the same features as the membrane-associated array (Fig.1C). However, the lattice constant was significantly increased by 1.1 nm to 13.6 +/- 0.1 nm (five determinations) which was

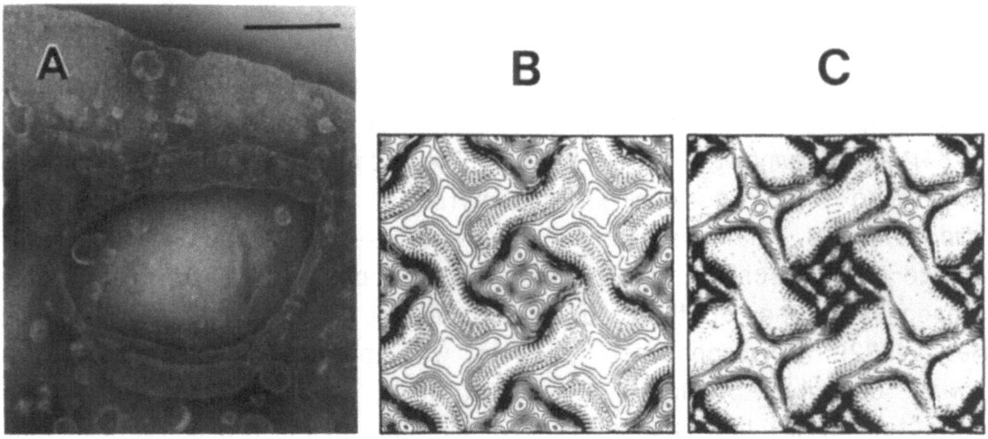

Fig.1. Electron micrographs of the S-layer of *A. vinelandii* negatively stained with ammonium molybdate. (A) S-layer associated with the cell wall. Bar = 200 nm. (B) Projection structures of the *A. vinelandii* S-layer obtained by correlation averaging of a number of electron micrographs containing areas of the cell-wall-associated S-layer and (C) S-layer fragments dislodged from the cell wall

associated with a 16^{0} clockwise rotation of the central tetrameric subunit. Assuming the stain-filled areas to be gaps in the S-layer, in terms of porosity, a transition from a "closed" to a more "open" structure was evident.

Three-Dimensional Structure of the *A. vinelandii* S-Layer

Although it may have been preferable to reconstruct the 3-D structure of the membrane-associated S-layer, the holes in the ghosts were generally too small to record images at the very high tilt angles necessary for a reliable 3-D reconstruction. Therefore the 3-D structure was reconstructed using negatively stained S-layer fragments free of the outer membrane support. Nineteen projections with actual tilt angles between 0 and 80.6^{0} were used for the 3-D reconstruction. The distribution of data in Fourier space showed heavy filling to a resolution of 2 nm with lighter filling to a resolution of 1.7 nm.

The 3-D model of the *A. vinelandii* S-layer is presented in Fig.2 as computer-generated views of the two surfaces of the S-layer. These surface-shaded representations give the impression that the central tetrameric subunit of the projection structure is funnel-shaped and flattened at the apex. These subunits, situated at one fourfold symmetry axis are interconnected at the other fourfold symmetry axis to form cruciform-linking structures. Because freeze-etch replicas of whole cells exhibit subunits separated by a distance of 12.5 nm (Bingle et al. 1987) in order to be consistent with the three-dimensional model, these subunits must represent either the cruciform-linking structures or the apex of the funnel-shaped subunits. A relief reconstruction of freeze-etched cells revealed tetrameric structures whose orientation of the linking arms of the cruciform subunits suggested that the cruciform-linking structures were situated at the outer layer surface. The apex of funnel-shaped tetramer, therefore, was in contact with the underlying outer membrane. The thickness of the layer, as suggested by the 3-D reconstruction is 5 nm.

The gaps in the S-layer are basically parallelogram-shaped in projection (11 nm x 3.5 nm, max.), arranged in simple cobblestone-like pattern and apparently open down to the outer membrane. If the 3-D model is forced to contain almost 90% of the expected volume, small features (about 1 nm thick) near the central layers of the linking arms are obtained connecting neighbouring arms. Probably the gaps are not completely open but contain small protein domains.

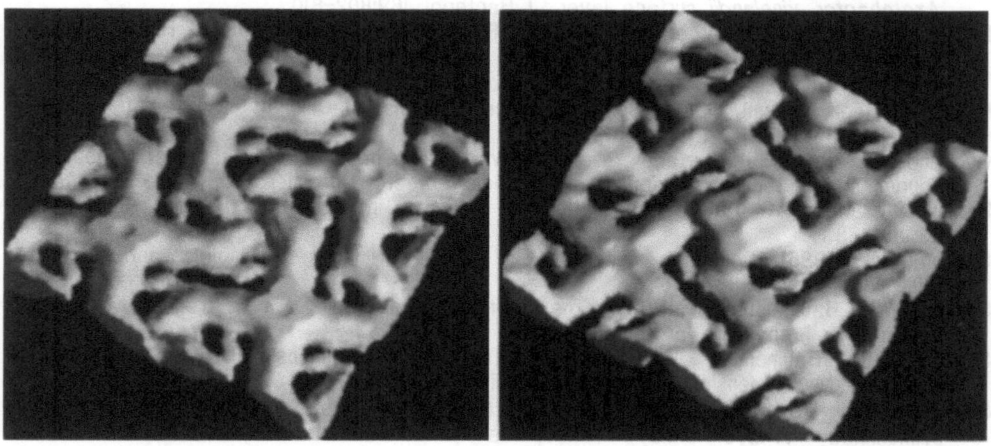

Fig.2. Computer-generated view of the two surfaces of the 3-D reconstruction of the *A. vinelandii* S-layer using a surface shading program. The funnel-shaped subunit is located in the center of the figure surrounded by the cruciform-linking structures. Left: outer surface of the S-layer. Right: inner surface of the S-layer. About 90% of the expected volume is included in the model. Dimensions of each image: 27 by 27 nm

In this study a small but significant alteration in the structure of the S-layer was noted upon removal from the cell wall. This was characterized by a rotation of the funnel-shaped subunit, an increase in the lattice constant and a transition, in terms of porosity, from a closed to a more open structure, similar to that observed by Stewart et al. (1986). Assuming the S-layer to be a closed bag-shaped structure and all the unit cells to be converted to the open conformation, the increase in the lattice constant would translate into an increase in volume enclosed by the S-layer bag of approximately 25%. This would require the S-layers of *A. salmonicida* and *A. vinelandii* to have a distance of 35 nm and 90 nm, respectively, from the outer membrane. As yet there is no evidence for a genuine local detachment of the S-layer in either case.

References

Baumeister W, Barth M, Hegerl R, Guckenberger R, Hahn M, Saxton WO (1986a) Three-dimensional structure of the regular surface layer (HPI) layer of *Deinococcus radiodurans.* J Mol Biol 187:241-253

Baumeister W, Guckenberger R, Engelhardt H, Woodcock CLF (1986b) Metal shadowing and decoration in electron microscopy of biological macromolecules. Ann N Y Acad Sci 483:57-76

Bingle WH, Doran JL, Page WJ (1984) Regular surface layer of *Azotobacter vinelandii.* J Bacteriol 159:251-259

Bingle WH, Whippey PW, Doran JL, Murray RGE, Page WJ (1987) Structure of the *Azotobacter vinelandii* surface layer. J Bacteriol 169:802-810

Engelhardt H, Guckenberger R, Hegerl R, Baumeister W (1985) High resolution shadowing of freeze-dried bacterial photosynthetic membranes: multivariate statistical analysis and surface relief reconstruction. Ultramicroscopy 16:395-410

Frank J, Heel M van (1982) Correspondence analysis of aligned images of biological particles. J Mol Biol 161:134-137

Guckenberger R (1985) Surface reliefs derived from heavy-metal-shadowed specimens. Fourier space techniques applied to periodic objects. Ultramicroscopy 16:357-370

Robish SA, Marr AG (1962) Localization of enzymes in *Azotobacter agilis.* J Bacteriol 83:158-168

Saxton WO (1985) Computer generation of shaded images of solids and surfaces. Ultramicroscopy 16:387-394

Saxton WO, Baumeister W (1982) The correlation averaging of a regularly arranged bacterial cell envelope protein. J Microsc 127:127-138

Saxton WO, Baumeister W, Hahn M (1984) Three-dimensional reconstruction of imperfect two-dimensional crystals. Ultramicroscopy 13:57-70

Stewart MJ, Beveridge TJ, Trust TJ (1986) Two patterns in the *Aeromonas salmonicida* A-layer may reflect a structural transformation that alters permeability. J Bacteriol 166:120-127

Structure and Assembly of the Oblique S-Layer Lattice of *Bacillus stearothermophilus* Strain NRS 2004/3a

D. Pum, P. Messner and U.B. Sleytr

Zentrum für Ultrastrukturforschung
Universität für Bodenkultur
Gregor Mendelstr. 33
A-1180 Wien
AUSTRIA

The ultrastructure and self-assembly products of the S-layer of *Bacillus stearothermophilus* strain NRS 2004/3a were investigated by electron microscopy using negatively stained and freeze-dried/unidirectionally shadowed preparations (Messner et al. 1986).

Depending on the dialysis conditions a broad spectrum of different S-layer self assembly products was found as shown in Fig.1. These could be separated into flat sheets and cylinders of different sizes. Statistical analysis of the projected diameters of the collapsed cylinders in negatively stained preparations lead to a classification into small (70 nm and 100 nm), medium (500 nm-600 nm) and large sized cylinders (ca. 1000 nm). In addition, small cylinders showed a different stain intensity in comparison to the medium and large cylinders.

By optical diffractometry it was found that small cylinders consisted only of monolayers while medium and large cylinders as well as flat sheets were composed of double layers. Finally the small cylinders were found in two populations to differ in their diameters (see above) and their lattice orientation in relation to the cylinder axis. The inherent curving tendency was found to be orthogonal for both types. Optical diffraction analysis of negatively stained images of cylindrical and sheet-like double layer assembly products was used to determine the rotational shift for the two layers. Five different superposition types with typical Moire patterns in the negatively stained images were found. The lateral alignment could be deduced from the position of absent or highlighted reflections in the diffraction pattern. The results could be confirmed by digital methods which were used for the subsequent reconstruction of the monolayer (see below). All double layers revealed a back-to-back orientation of the two constituent layers. Type I had the short base vectors of the oblique unit cell in common and was found only in flat sheets. In Type II the long base vectors of both oblique unit cells were parallel but shifted by half a repeat of the corresponding distance. Large and medium as well as flat sheets were found for this type. In Type III only the (2,-1) lattice lines were parallel but again shifted half a repeat of the spacing. Flat sheets and medium-sized cylinders were found for this superposition type. Type IV had the short diagonals but shifted half a repeat, and Type V, with the long diagonals again shifted half a repeat in parallel. Type II was found to be the most abundant type of assembly products followed by Type IV and Type III. Type I and Type V were only rarely detected. Exact measurement of the lattice parameters and t-tests (1% error probability) revealed significant differences in the interaxial angle amongst different assembly types (Type I, II : ca. 78^{0}, Type II, IV, V : ca. 81^{0}) while the base vector lengths (a = 9.4 nm, b = 11.6 nm) remained the same. The Moiré pattern of Type II is shown in Fig.2 and the corresponding diffraction pattern in Fig.3. The mirror symmetric orientation of the two reciprocal lattices follows from the back-to-back orientation of the two constituent layers. The Delta filter was used to obtain a one-sided filtered image of a negatively stained double layer. The reconstruction with a resolution of 1.9 nm is shown in Fig.4. Well-defined stain/protein boundaries with a steep gradient in density can be seen. The bulk of the protein, which is a dimer with characteristic handedness is centered around the central two-fold axis of the oblique (p2) unit cell. The high regularity of the p2-lattice, which becomes evident by comparing p1- and p2-averages, could be quantified by mean phase residual values of 7^{0} - 8^{0} for all reconstructions up to the resolution limit.

A three-dimensional model of the S-layer structure was calculated by combining tilt series in the angular range from -60^{0} to $+60^{0}$ with Fourier methods as described by Amos et al. (1982). Due to the low symmetry of the space group p2, 72 tilted views were used. The resolution of the 3D-model is 2 nm. Fig.5 shows a top-on view and Fig.6 a bottom-on view of the structure.

Freeze-dried and heavy metal shadowed preparations (Pt/C and Ta/W) were also reconstructed using correlation averaging methods as described by Saxton and Baumeister (1982). Depending on the azimuth of the shadowing source, three prominent furrows on the outer surface of the

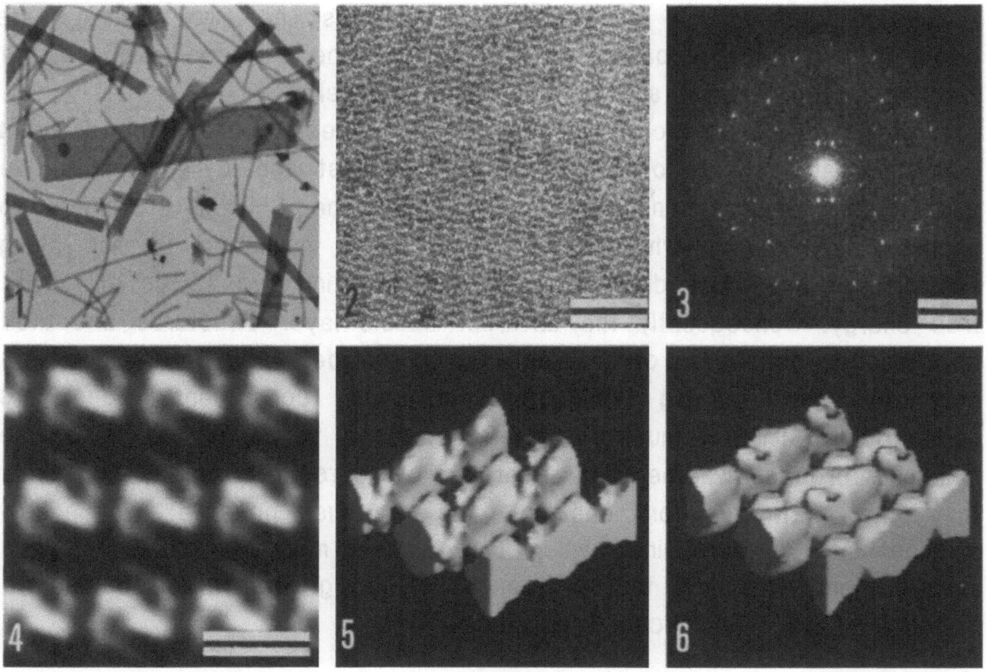

Fig.1. A mixture of small, medium and large sized cylindrical self assembly products in a negatively stained preparation. Bar = 1 µm
Fig.2. Moiré pattern of the negatively stained double layer of Type III. Bar = 100nm
Fig.3 Diffraction pattern to Fig.2. The mirror-symmetrical orientation of the two reciprocal lattices show the back-to-back orientation of the monolayers in the double layer. Bar = (1/15) nm^{-1}
Fig.4. 'One-sided' Delta filtered image reconstruction of the negatively stained monolayer structure of the double layer of Type III. Bar = 10nm
Fig.5. Three-dimensional model of the S-layer structure in a top-on view. The model shows a central subunit with its dimer and the nearest neighbouring subunits
Fig.6. Same as Fig.5 but in a bottom-on view

S-layer could be visualized, as already seen in freeze-etched images of intact cells (Messner et al. 1986).

Labelling with polycationic ferritin (PCF) which is a positively charged marker, demonstrated differences in the charge distribution between inner and outer surface of the monolayer. PCF bound neither on the outer surface of the S-layer on the intact cell nor on double layers in general. But PCF bound on the inner side of the small cylinders and on small monolayer areas in double layer assemblies where the double layer was incomplete. This was found in negatively stained and freeze-dried/unidirectionally shadowed images. These findings lead to the conclusion that for all double layer assemblies the two S-layers are oriented in a back-to-back orientation where the surfaces exposed are identical to the S-layer surface seen on intact cells.

Two types of monolayer cylinders have been found with different diameters and orientation of the lattice against the cylinder axis. We explain this phenomenon with the ability of the S-layer protomers to generate monolayers which have the inherent inclination to roll along two distinct axes with different curvature reflecting two states of lowest free energy. Consequently the different double layer structures and their appearances as sheets or differently sized cylinders can be explained by combining the curving tendencies of the constituent monolayers geometrically. Thus it must be determined at a very early stage of the assembly process whether flat sheets or cylinders are formed. It is likely that the pathway for one of the various possible free-energy arrangements is determined when only a few morphological units have aligned themselves and that differences in the growth rates of the two monolayers are of significant importance.

References

Amos LA, Henderson R, Unwin PNT (1982) Three-dimensional structure determination by electron microscopy of two-dimensional crystals. Prog Biophys Mol Biol 39:183-231

Messner P, Pum D, Sleytr UB (1986) Characterization of the ultrastructure and the self-assembly of the surface layer of *Bacillus stearothermophilus* strain NRS 2004/3a. J Ultrastruct Mol Struct Res 97:73-88

Saxton WO, Baumeister W (1982) The correlation averaging of a regularly arranged bacterial cell envelope protein. J Microsc 127:127-138

Charge Distribution of S-Layers and Importance of Charged Groups for Morphogenesis and Function

M. Sára and U.B. Sleytr

Zentrum für Ultrastrukturforschung
Universität für Bodenkultur
Gregor-Mendel Str. 33
A-1180 Wien
AUSTRIA

Bacterial cell surfaces are generally recognized as having a net negative charge and isolelectric point in the range of 3 (Wicken 1985). By contrast, the S-layer carrying highly thermophilic Bacillaceae *Bacillus stearothermophilus* NRS 1536/3c, *Clostridium thermosaccharolyticum* D120-70 and *C. thermohydrosulfuricum* L111-69 were not capable of binding polycationized ferritin (PCF) (Sára et al. 1988). This indicates that the outer face of the S-layer lattice does not exhibit a net negative charge. PCF labelling of isolated S-layers revealed that the inner face of the S-layer lattice adhering to the peptidoglycan-containing layer in whole cells is net negatively charged. Thus, as far as investigated, S-layers of the thermophilic Bacillaceae represent highly polar, anisotropic membranes. The dipole characteristics of the protomers may be required for a correct orientation during local insertion in the course of lattice growth (Sleytr and Messner 1983).

Crosslinking of the S-layer lattices with glutaraldehyde and subsequent carboxyl-specific modification reactions confirmed the presence of free amino and carboxyl groups on the outer face of the S-layer lattice of *B. stearothermophilus* NRS 1536/3c and the two clostridia. Glutaraldehyde-treated whole cells were capable of PCF binding, indicating that free amino groups exposed on the outer S-layer face had been blocked. After activation of the carboxyl groups with carbodiimides and reaction with glycine methyl ester as nucleophile the glutaraldehyde-treated S-layer surfaces lost the ability to bind PCF.

The arrangement of the surface located charged groups was investigated by using amino-specific, homobifunctional crosslinkers and carbodiimides (Peters and Richards 1977). Upon treatment of the square

S-layer lattice of *B. stearothermophilus* NRS 1536/3c with imidoesters of various crosslinking spans and the N-hydroxy-succinimide ester of suberic acid (NHS), no protein bands could be detected on SDS gels (Sára and Sleytr 1987a) indicating the introduction of covalent linkages between the adjacent subunits. These results showed that at least two free amino groups must be exposed on the surface of each protomer, occurring at a distance of approximately 1.0 nm to those of the adjacent subunits. In the case of the two clostridia the S-layer protein remained at least partly SDS-extractable upon treatment with dimethyladipimidate (DMA; crosslinking span 0.9 nm) or dimethylsuberimidate (DMS; crosslinking span 1.1 nm). Besides the monomeric S-layer protein bands, reaction products with higher apparent molecular weights were present on SDS gels (Table 1) being attributed to the formation of dimers. The different but distinct migration properties of these reaction products on SDS gels can be explained by a structural heterogeneity due to the reaction of amino groups located at different lattice positions. Dimethylpimelinimidate (DMP; crosslinking span 1.0 nm, reacting at pH 8.5) and NHS (crosslinking span 1.1 nm, reacting at pH 7.2) could crosslink the subunits in the S-layer lattices of both clostridia.

Upon treatment of the S-layer lattices with the carboxyl-specific carbodiimides, only the S-layer protein of *C. thermosaccharolyticum* D120-70 remained SDS-extractable. It is evident that it is only in such S-layer lattices where crosslinking with carbodiimides occurs that carboxyl and amino groups originating from adjacent protomers are arranged in close proximity. Such an arrangement will lead to direct electrostatic interactions between the constituent subunits in the native arrays.

Chemical modification reactions performed on S-layer lattices included amidation of carboxyl groups, amidination of amino groups with ethylacetimidate, N-acetylation and N-succinylation. For amidation the carboxyl groups were activated with carbodiimides and converted into neutral groups by using glycine methyl ester as nucleophile. N-succinylation led to a complete disintegration of all S-layer lattices. The released subunits were water-soluble, showing 7% unmodified amino groups. During N-acetylation approximately 90% of the S-layer protein was liberated. The released subunits, showing 7% unmodified amino groups, had lost the ability to reassemble or to reaggregate. In the case of *B. stearothermophilus* NRS 1536/3c, the acetylated protomers remaining attached to the underlying cell envelope layer possessed 17%

unmodified amino groups. Upon extraction from the rigid cell wall layer with chaotropic agents, the acetylated subunits reaggregated into amorphous material. Amidination retained the native charge distribution but enlarged the electropositive residue. Amidinated protomers showing only 7% unmodified amino groups were still capable of reassembling into regularly structured lattices identical to those obtained with native S-layer protein. Amidation of the carboxyl groups led to the loss of the regular structure of the hexagonal S-layer lattice of *C. thermohydrosulfuricum* L111-69, although the amidated protomers remained attached to the peptidoglycan-containing layer. A conversion of the carboxyl

Table 1. Crosslinking studies on S-layer lattices of two clostridia. Apparent molecular weights of SDS-extractable reaction products after treatment with dimethyladipimidate (DMA), dimethylpimelinimidate (DMP), dimethylsuberimidate (DMS), N-hydroxy succinimide ester of suberic acid (NHS) or 1-ethyl-3,3 (dimethylaminopropyl)carbodiimide (EDC).

Crosslinking agent, crosslinking span	Apparent molecular weights on SDS gels *C. thermosaccharolyticum* D120-70	*C. thermohydrosulfuricum* L111-69
Native	115,000	120,000
DMA (0.9 nm)	190,000 200,000	120,000 170,000 180,000 200,000
DMP (1.0 nm)	no protein bands	no protein bands
DMS (1.1 nm)	115,000 190,000 200,000	120,000 170,000 180,000 200,000
NHS (1.1 nm)	no protein bands	no protein bands
EDC (0)	115,000	no protein bands

groups into neutral groups obviously led to the breakdown of electrostatic interactions between the adjacent protomers. Isolated carboxy-amidated subunits of *B. stearothermophilus* NRS 1536/3c had lost the ability to reassemble into regularly structured lattices.

The modification studies confirmed that the native charge distribution in the protomers is required for maintaining the structural integrity of S-layer lattices, as well as for enabling an unhindered self-assembly process. The presence of surface located charged groups strongly suggests that S-layers favour adhesive activities of the living cell, and can also act as ion exchange resins which can attract and bind organic and inorganic compounds in the area immediately surrounding the cell. Since S-layers represent molecular sieves (Sára and Sleytr 1987b), charged groups, particularly in the pore areas, could significantly contribute to the exchange of nutrients and metabolites between the cell and its environment. Moreover, the immobilization of toxic or hostile materials always present in natural environments by charged groups may also be suggested (Beveridge 1981).

References

Beveridge TJ (1981) Ultrastructure, chemistry and function of the bacterial cell wall. Int Rev Cytol 72:229-317

Peters K, Richards FM (1977) Chemical crosslinking reagents and problems in studies of membrane structure. Annu Rev Biochem 46:523-551

Sára M, Sleytr UB (1987a) Charge distribution on the S-layer of *Bacillus stearothermophilus* NRS 1536/3c and importance of charged groups for morphogenesis and function. J Bacteriol 169:2804-2809

Sára M, Sleytr UB (1987b) Molecular sieving through S-layers of *Bacillus stearothermophilus* strains. J Bacteriol 169:4092-4098

Sára M, Kalsner I, Sleytr UB (1988) Surface properties of the S-layer of *Clostridium thermosaccharolyticum* D120-70 and *Clostridium thermohydrosulfuricum* L111-69. Arch Microbiol (in press)

Sleytr UB, Messner P (1983) Crystalline surface layers on bacteria. Annu Rev Microbiol 37:311-339

Wicken AJ (1985) Bacterial cell walls and surfaces. In: Savage DC, Fletcher M (eds) Bacterial adhesion. Mechanisms and significance. Plenum, New York, pp 45-70

Localized Insertion of New S-Layer During Growth of *Bacillus stearothermophilus* Strain PV72

K. Gruber and U.B. Sleytr
Zentrum für Ultrastrukturforschung
Universität für Bodenkultur
Gregor Mendel Str. 33
A-1180 Wien
AUSTRIA

Bacillus stearothermophilus PV72 carries a crystalline surface layer (S-layer) with hexagonal symmetry (Sleytr and Messner 1983). Sites of insertion of newly synthesized subunits into the regular lattice during cell growth have been determined using the indirect fluorescent antibody technique (Hughes and Stokes 1971) and the protein A/colloidal gold marker system (Mühlpfort 1982, Smit and Agabian 1982). In both systems, cells from a logarithmic growing culture have been incubated with anti-S-layer serum in order to completely cover the cells and then reincubated in fresh medium. Samples were extracted at intervals and reacted with fluoresceinisothiocyanate conjugated goat-anti-rabbit-IgG or protein A/colloidal gold conjugate. S-layer material incorporated into the lattice after reincubation had no IgG attached to it and therefore did not bind the marker.

Results obtained at light microscope level are shown in Fig.1. The appearance of unstained areas at distinct regions on the cell surface in the course of time represents the distribution of incorporation sites of newly synthesized S-layer material into the existing surface layer. In the 2D projection of the cell surface, the unstained areas are particularly clearly seen at the edges of the rods. The results obtained at electron microscope level (Fig.2) - the dark central region in this series of micrographs is due to the shrunken protoplast which retracted from the cell wall, confirming the results obtained with the fluorescent antibody technique and thus the following conclusions were drawn:

- at the cylindrical part of the cell, new S-layer material is deposited in spiral-like bands at a pitch angle related to the orientation of the lattice vectors of the crystalline array
- little or no S-layer material is inserted into existing cell poles

- septal regions and the subsequently formed new cell poles are covered with new S-layer protein

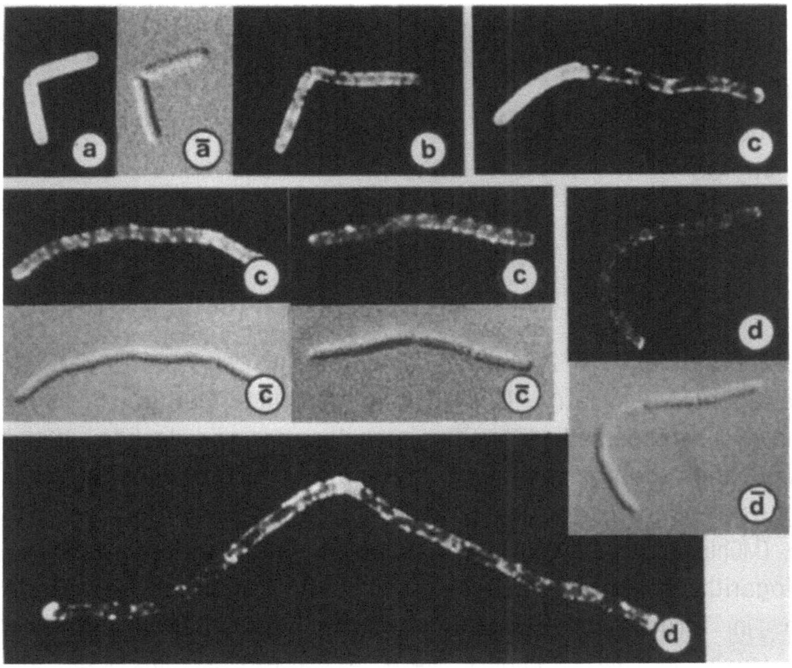

Fig.1. Labelling of the old S-layer of *Bacillus stearothermophilus* PV72 with the indirect fluorescent antibody technique. (a,ā) at 0 time; (b) after 15 min; (c,c̄) after 30 min; (d,d̄) after 60 min of growth. (a,b,c,d) fluorescence illumination, (ā,c̄,d̄) phase interference contrast; x 2100; generation time approximately 20 min

As shown in the schematic drawing (Fig.3) the geometry and distribution of the growth areas can be explained by two mechanisms: firstly, the entropy-driven incorporation of new S-layer monomers most likely occurs along sliding dislocations, whereby the orientation of the dislocations will be determined by the alignment of the lattice on the cylindrical part of the cell. As can be seen in freeze-etching replicas (Fig.3a), strain PV 72 has a hexagonally ordered S-layer with most crystalline domains oriented in such a way that one symmetry axis runs parallel to the longitudinal axis of the cell (Sleytr et al. 1982). As a consequence of this sliding dislocations as potential S-layer incorporation sites would likely follow the two symmetry axes in the hexagonal lattice which assume an approximate 60° angle with respect to the longitudinal axis

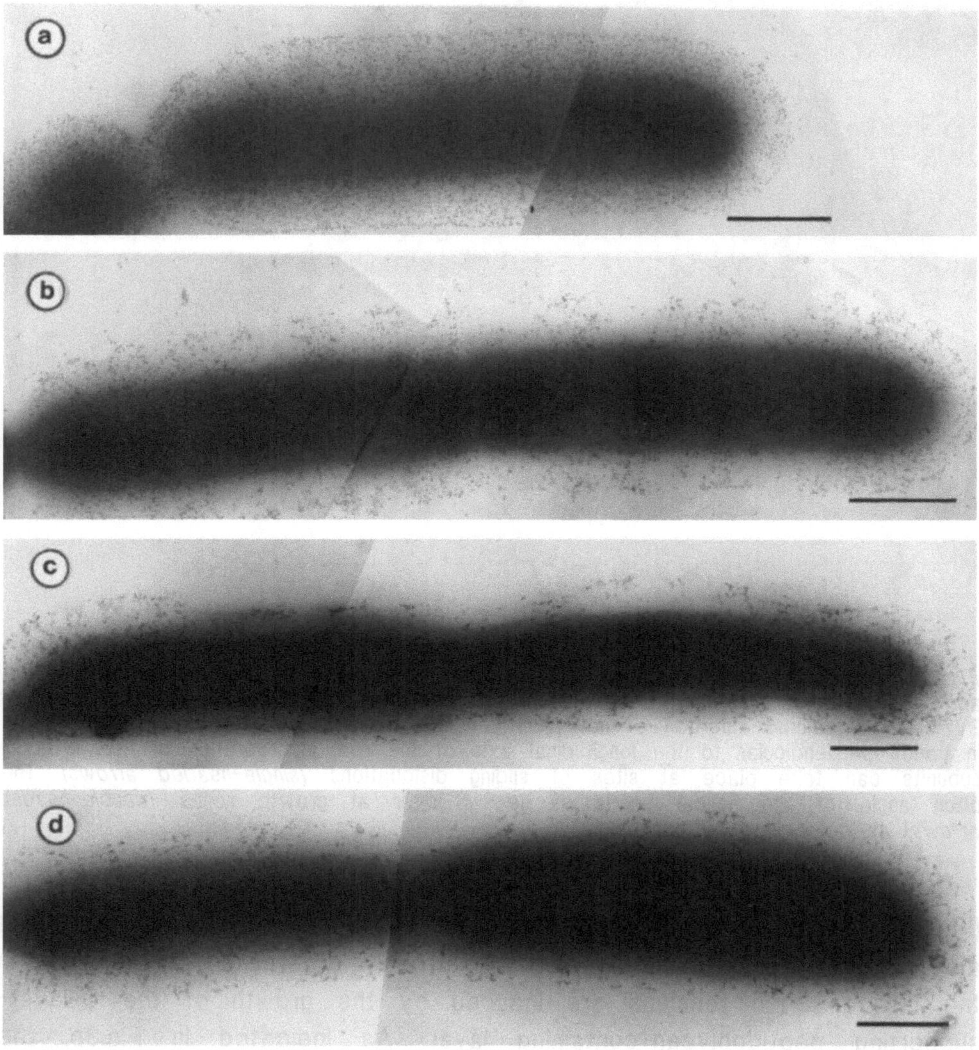

Fig.2. Labelling of the old S-layer of *Bacillus stearothermophilus* PV72 with the protein A/colloidal gold technique. (a) at 0 time; (b) after 30 min; (c,d) after 60 min of growth; bars represent 0.5 μm

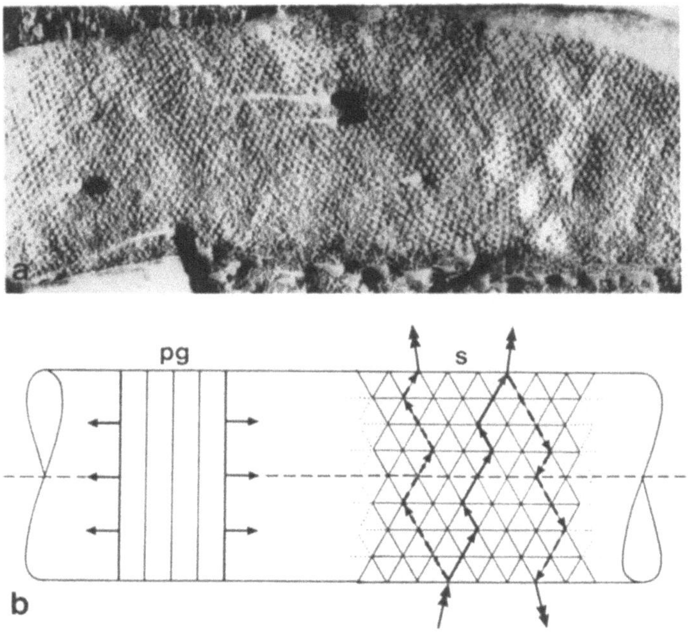

Fig.3. Diagrammatic representation (a) of the expansion of the peptidoglycan *(pg)* and the S-layer *(s)* and freeze-etching preparation of ***Bacillus stearothermophilus*** PV72 (b). Localized insertion of new peptidoglycan material can be expected to occur in bands perpendicular to the longitudinal axis of the rod. Incorporation of new S-layer subunits can take place at sites of sliding dislocations *(single-headed arrows).* The pitch angle of the helical bands of new S-layer at growth zones *(double-headed arrows)* as determined by peptidoglycan

(Fig.3b). Secondly, the preferential incorporation of S-layer subunits along sliding dislocations will also be strongly determined by tensions in the lattice. Such tensions are induced by the growth of the S-layer-supporting peptidoglycan-containing layer. As indicated in Fig.3b, the glycan strands of the murein sacculus in Gram-positive rod-shaped bacteria most likely run perpendicular to the long axis of the cell (Hughes and Stokes 1971, Pooley et al. 1978).

It appears that in the strain examined the boundary line between the old and new S-layer preferentially runs at an angle somewhat between the directions of the lattice vectors and the assumed direction of growth of the murein sacculus. Consequently, S-layer growth can be expected to take a step-wise course, involving incorporation of new subunits along sliding dislocation and jumps into adjacent rows.

The observation that newly synthesized S-layer bands in *Bacillus sphaericus* run perpendicular to the long axis of the cell (Howard et al. 1982) has accordingly been explained by the arrangement of the axis of symmetry in the tetragonal lattice. In contrast, the random growth of the hexagonal S-layer of the Gram-negative *Caulobacter crescentus* has been attributed to the nature of the outer membrane and the excretion of S-layer protein over the Bayer junctions.

References

Howard LV, Dalton DD, McCoubry WK (1982) Expansion of the tetragonally arrayed cell wall protein layer during growth of *Bacillus sphaericus*. J Bacteriol 149:748-757

Hughes RC, Stokes E (1971) Cell wall growth in *Bacillus licheniformis* followed by immunofluorescence with mucopeptide-specific anti-serum. J Bacteriol 113:1491-1504

Mühlpfort H (1982) The preparation of colloidal gold particles using tannic acid as an additional reducing agent. Experientia 38:1127-1128

Pooley HM, Schlaeppi J-M, Karamata D (1978) Localized insertion of new cell wall in *Bacillus subtilis*. Nature (London) 274:264-266

Sleytr UB, Messner P (1983) Crystalline surface layers on bacteria. Annu Rev Microbiol 37:311-339

Sleytr UB, Messner P, Schiske P, Pum D (1982) Periodic surface structures on prokaryotic cells. Proc 10th Int Congr Electron Microsc, Hamburg, 3:1-8

Smit J, Agabian N (1982) Cell surface patterning and morphogenesis: biogenesis of a periodic surface array during *Caulobacter* development. J Cell Biol 95:41-49

Crystalline Surface Layers in Anaerobic Bacteria Isolated from a Patient with Secretory Otitis Media and a Draining Ear

K. Lounatmaa[1], H. Jousimies-Somer[2], R. Grénman[3] and A. Rintala[3]

[1]Departments of Electron Microscopy and General Microbiology
University of Helsinki
Mannerheimintie 172
[2]The Anaerobe Reference Unit, National Public Health Institute
[3]The Helsinki University Central Hospital
1st Department of Surgery, Cleft Center
SF-00280 Helsinki, FINLAND

Secretory otitis media is globally a very widespread complication in young children with unrepaired cleft palate (Koivuniemi 1982). The malfunction of the Eustachian tube can cause the negative pressure and accumulation of fluid in the middle ear cavity. Secretory otitis media was earlier thought to be a "sterile" process, but recently numerous pathogenic bacteria have been isolated (Jousimies-Somer et al. 1986). The middle ear drainage through the ventilation tube can be due to the secondary infection after insertion of the tubes (Vogelsang and Birck 1984). The bacteriology of "draining ears" and the virulence factors of individual species is not yet understood.

Methods

Isolation and identification of bacterial species. The samples were immersed in a small volume of thioglycollate broth. The homogenized material was inoculated onto the blood and chocolate agar, supplemented Brucella and PEA (phenylethylalcohol) agars and enriched in thioglycollate broth (Sutter et al. 1980). The anaerobic cultures were incubated for 48 - 96 h in Gas-Pak jars (BBL, Cockeysville, MD) and isolates identified by determining reactions using PRAS (prereduced, anaerobically sterilized) biochemicals (Holdeman et al. 1977), the profile of performed enzymes by the Rap ID Ana System (Innovative Diagnostic System Inc., Atlanta, GA) and other characteristics by methods outlined in the manuals of Virginia Polytechnic Institute (VPI) and Wadsworth Anaerobe Laboratory.

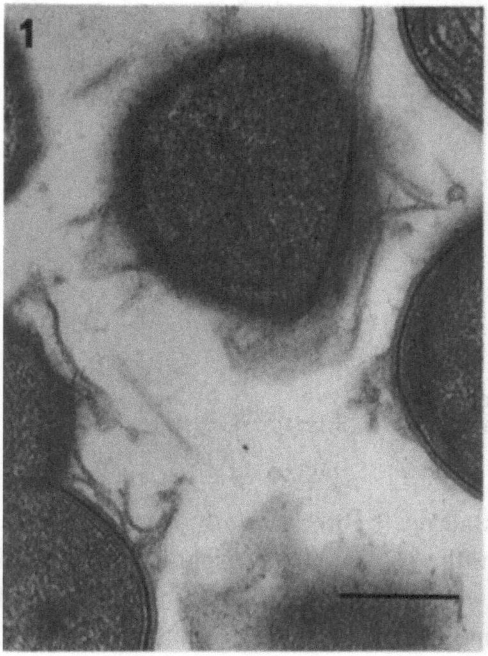

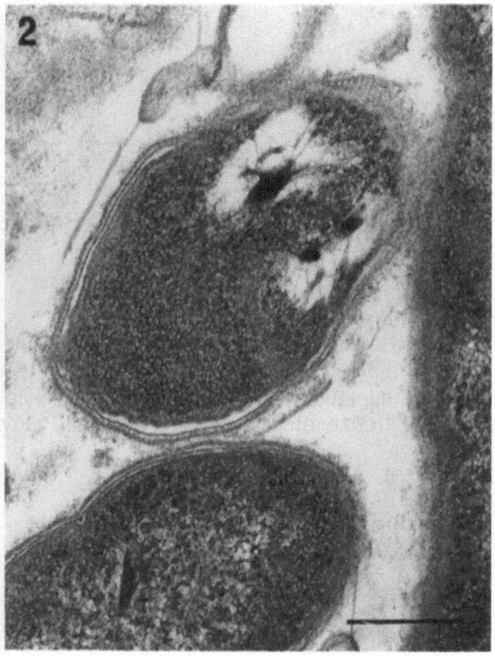

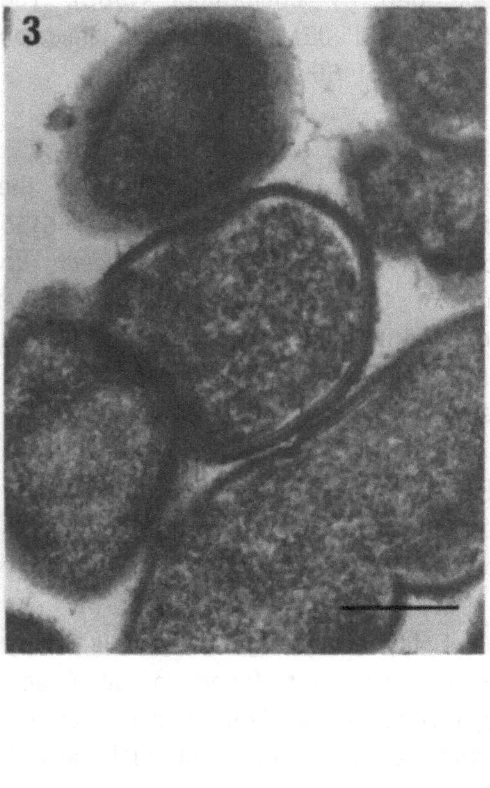

Fig.1. Thin-sectioned cells of *Peptostreptococcus magnus* AHC 5155. The S-layer is peeling off from the cell wall. Bar in all the micrographs = 0.2 µm

Fig.2. Thin-sectioned cells of *Eubacterium lentum* AHP 6099 with releasing S-layer fragments

Fig.3. Thin-sectioned cells of *Eubacterium* sp. AHN 990 with very electron dense S-layers on the cells

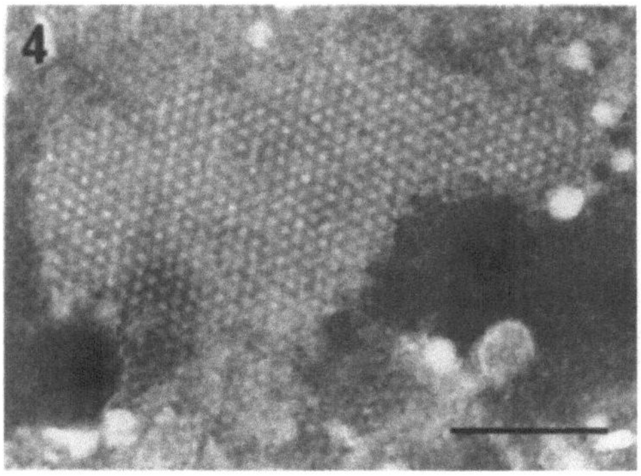

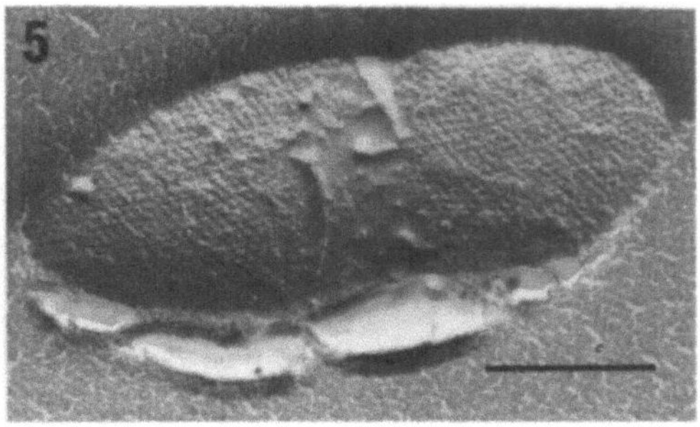

Fig.4. A negatively stained S-layer fragment of AHN 990
Fig.5. A freeze-etched cell of AHN 990 covered by the S-layer

Electron microscopy. For negative staining bacteria were suspended in double distilled water and stained with 1% (w/v) phosphotungstic acid (pH 6.5). Samples for thin sections were prefixed in glutaraldehyde and post-stained with uranyl acetate and lead citrate (Lounatmaa 1985). Freeze-etching was performed as described by Lounatmaa (1985). The micrographs were taken with JEM-100CX and JEM-1200EX transmission electron microscopes at an operating voltage of 60 kV.

Results and Discussion

The S-layer was found to be the outermost cell wall layer in the following Gram-positive species: *Peptostreptococcus magnus* AHC 5155 (Fig.1), *Eubacterium lentum* AHP 6099 (Fig.2) and *Eubacterium* sp. AHN 990 (Figs.3-5). The ultrastructure of the cell wall of *Eubacterium* sp. AHN 990 is described here in more detail: the paracrystalline protein layer is very electron dense and seems to be much thicker than the cell wall (Fig.3). The hexagonal arrangement of the subunits of the array is slightly visible in thin-sectioned cells (Fig.3) but very prominent in a negatively stained S-layer fragment (Fig.4) as well as in a freeze-etched cell (Fig.5). Using crystallographic techniques, Hovmöller and coworkers (this vol) have determined the lattice constant to be 15.6 nm and have done both two- and three-dimensional reconstructions of this layer.

All the anaerobic bacteria with the S-layer investigated in this study were Gram-positive, although the S-layer was not found in all Gram-positive isolates and some of them still remained unclear in this respect. In addition, all the S-layers had hexagonal orientation, although among the Gram-positive bacteria tetragonal was found more often (Sleytr and Messner 1983).

The significance of these S-layer bacteria in the pathogenesis of otitis and persistence of infection in our patient is difficult to explain and remains to be solved.

References

Holdeman LV, Cato EP, Moore WEC (eds) (1977) Anaerobic laboratory manual, 4th edn. Virginia Polytech Inst State Univ, Blacksburg

Jousimies-Somer H, Grénman R, Rintala A (1986) Bacteriological investigation of secretory otitis media in children with cleft palate. Scand J Plast Reconstr Surg 20:297-302

Koivuniemi P (1982) Secretory otitis media and its management in children with cleft palate. Duodecim 98:995-998

Lounatmaa K (1985) Electron microscopic methods for the study of bacterial surface structures. In: Korhonen TK, Dawes EA, Mäkelä PH (eds) Enterobacterial surface antigens: methods for molecular characterization. Elsevier, Amsterdam, pp 243-261

Sleytr UB, Messner P (1983) Crystalline surface layers on bacteria. Annu Rev Microbiol 37:311-339

Sutter VL, Citron DM, Finegold SM (1980) Wadsworth anaerobic bacteriology manual, 3rd edn. Mosby, St Louis

Vogelsang MV, Birck HG (1984) Ventilation tubes in the pediatric population. In: Lim DJ, Bluestone CD, Klein JO, Nelson JD (eds) Recent advances in otitis media with effusion. Decker, Philadelphia, pp 306-308

Occurrence and Characterization of S-Layers in Oral Bacteria

H. Ranta[1], M. Haapasalo[1], E. Kerosuo[1,2] and K. Lounatmaa[2,3]

[1]Department of Cariology
[2]Department of Electron Microscopy and
[3]Department of General Microbiology
University of Helsinki
Mannerheimintie 172
SF-00280 Helsinki, FINLAND

The first observations of additional surface layers composed of regularly arranged protein subunits in oral bacteria were reported at the beginning of this decade (Kornman and Holt 1981, Lai et al. 1981, Ranta et al. 1983, Haapasalo et al. 1985). The function of S-layers is not known, but they may play a crucial role in complex interactions of cells with their environment. Interestingly, bacteria causing human infections have rarely been shown to possess an S-layer.

Structural knowledge about S-layers of bacteria of human origin is mainly based on two-dimensional projections. The three-dimensional structure of the S-layer with 21.5 nm and 7.7 nm unit cell spacings in *Bacteroides buccae* has been reported (Sjögren et al. 1985). In addition, cells of *B. buccae* have been shown to possess a hexagonally arranged periodic structure (COMP) in the outer membrane (Kerosuo et al. 1987). The structural relationship of these two cell wall components remains to be elucidated. In Table 1, the present knowledge of oral species with an S-layer is summarized.

Methods

Isolation of S-layers. Cells of *B. buccae* strain ES57 were vigorously pipetted through a narrow tip in PBS (0.85% NaCl, 0.15% phosphate, pH 7.2) and centrifuged (9800 *g,* 5 min). The supernatant was further centrifuged at 72,000 *g* for 1 h. The supernatant was frozen and lyophilized. The fractionation of polypeptides was achieved by fast protein liquid chromatography (FPLC). The anion exchange column Mono Q (HR 5/5) was equilibrated with 10 mM Tris, pH 7.5 and a linear gradient of 0-1 M NaCl was created over a total volume of 10 ml at a flow rate of 0.5 ml/min. The fractions were further examined by gel electrophoresis in the presence of SDS with 4% stacking gel and 15% running gel.

Immunization. The rabbits received three injections (popliteal lymph node immunization) containing 20 ug protein in Freunds complete adjuvant and were killed 2 months later.

Immunoelectron microscopy. Cells were prefixed with 3.5% paraformaldehyde and 0.05% glutaraldehyde in 0.1 M phosphate buffer, pH 7.2. After incubation with rabbit serum (1:50 dilution) the cells were washed with phosphate buffer. After treatments with goat-anti-rabbit-IgG, rabbit-anti-ferritin and horse ferritin, washing with phosphate buffer was repeated. Finally, the cells were fixed with 3% glutaraldehyde. The electron micrographs were taken with a JEM-1200EX electron microscope at 60 kV.

Table 1. Oral bacteria with an S-layer

Species/strain	Symmetry	COMP	Reference
Gram-negative			
Bacteroides buccae	NR	NR	Kornman and Holt 1981
	h	NR	Ranta et al. 1983
B. buccae	h	+	Kerosuo et al. 1987
B. heparinolyticus	NR	NR	Okuda et al. 1985
	h	-	Kerosuo et al.(this vol)
B. forsythus	t[a]	NR	Tanner et al. 1986
Wolinella recta	h	NR	Lai et al. 1981
Gram-positive			
ES4C[b]	t		Kerosuo et al. 1988

COMP = crystalline outer membrane protein(s)
h = hexagonal, t = tetragonal
+ = present, - = not detected
NR = not reported
[a] Described as orthogonal
[b] Preliminarily identified as *Eubacterium* sp.

Results and Discussion

Oral bacteria with S-layers in association with human infections have been most often isolated from periodontally diseased sites. *B. buccae* (Ranta et al. 1983) and a novel Gram-positive anaerobic rod ES4C (Kerosuo et al. this vol) have been found in apical periodontitis (infection of the jaw bone). Whether the S-layer is a true virulence factor is not known. The S-layer of *B. buccae* was readily detached from the cells, in contrast to the tetragonal S-layer of strain ES4C. The latter was found resistant to both mechanical and enzymatic treatments. Only after prolonged spheroplast formation could crude S-layer preparations be obtained from strain ES4C. The S-layer of *B. buccae* appeared to be composed of three major polypeptides (gel not shown). Two of them had an apparent molecular weight of 15 kDa in addition to the 70 kDa component. For immunization, the 70kDa polypeptide was used. Specific binding of the antiserum to the S-layer was visualized by ferritin labelling (Fig.1). The relation of the 16 kDa polypeptides to the S-layer remains to be investigated. Although a protease inhibitor (phenylmethylsulphonylfluoride) was added, the possibility of proteolytic degradation cannot be excluded.

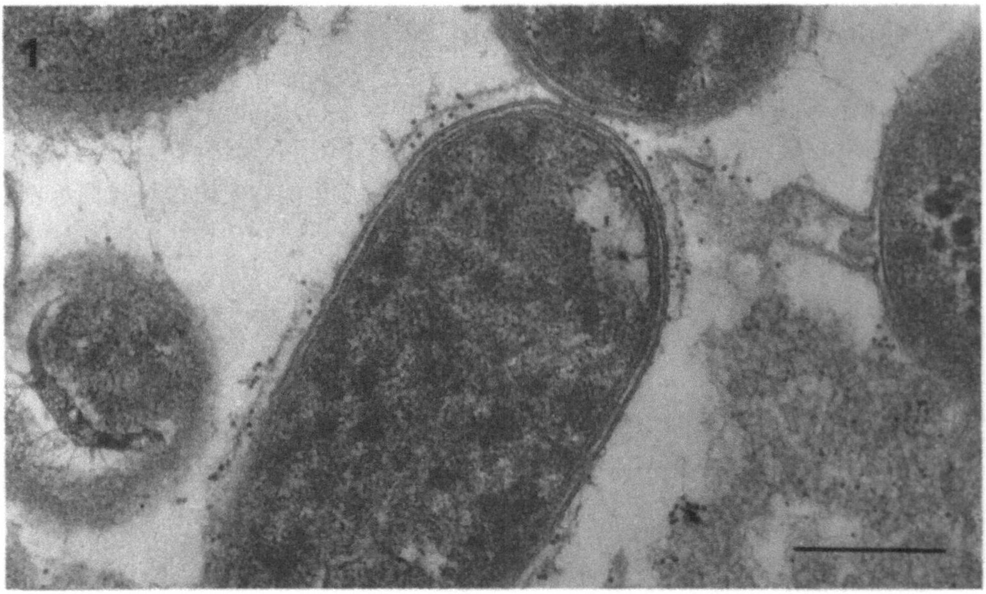

Fig.1. Binding of the antiserum to the S-layer (S) of *B. buccae* ES57 (ferritin labelling). Bar = 0.2 μm

References

Haapasalo M, Lounatmaa K, Ranta H, Shah H, Ranta K (1985) Ultrastructure of *Bacteroides capillus, B. buccae, B. pentosaceus, B. oris, B. oralis, B. veroralis,* and pentose sugar-fermenting *Bacteroides* sp. from humans with periapical osteitis: occurrence of external proteinaceous cell wall layer. Int J Syst Bacteriol 35:65-72

Kerosuo E, Haapasalo M, Ranta H, Lounatmaa K (1987) Hexagonal periodicity in the outer membrane of *Bacteroides buccae.* J Gen Microbiol 133:2217-2224

Kerosuo E, Haapasalo M, Lounatmaa K, Ranta H, Ranta K (1988) Ultrastructure of a novel anaerobic Gram-positive nonsporing rod from dental root canal. Scand J Dent Res (in press)

Kornman KS, Holt SC (1981) Physiological and ultrastructural characterization of a new *Bacteroides* species, *(Bacteroides capillus)* isolated from severe localized periodontitis. J Periodontal Res 16:542-555

Lai CH, Listgarten MA, Tanner ACR, Socransky SS (1981) Ultrastructure of *Bacteroides gracilis, Campylobacter concisus, Wolinella recta,* and *Eikenella corrodens,* all from humans with periodontal disease. Int J Syst Bacteriol 31:465-475

Okuda K, Kato T, Shiozu J, Takazoe I, Nakamura T (1985) *Bacteroides heparinolyticus* sp. nov. isolated from humans with periodontitis. Int J Syst Bacteriol 35:438-442

Ranta H, Lounatmaa K, Haapasalo M, Ranta K (1983) Isolation and ultrastructure of *Bacteroides* sp. with external cell-wall layer (S-layer) in periapical osteitis. Scand J Dent Res 91:458-464

Sjögren A, Hovmöller S, Farrants G, Ranta H, Haapasalo M, Ranta K, Lounatmaa K (1985) Structures of two different surface layers found in six *Bacteroides* strains. J Bacteriol 164:1278-1282

Tanner ACR, Listgarten MA, Ebersole JL, Strzempko MN (1986) *Bacteroides forsythus* sp. nov., a slow-growing, fusiform *Bacteroides* sp. from the human oral cavity. Int J Syst Bacteriol 36:213-221

Ultrastructural Comparison of *Bacteroides heparinolyticus* and *Bacteroides buccae*

E. Kerosuo[1,2], M. Haapasalo[1], K. Lounatmaa[2,3] and H. Ranta[1,4]

[1]Department of Cariology
[2]Department of Electron Microscopy and
[3]Department of General Microbiology
University of Helsinki
Mannerheimintie 172
[4]National Public Health Institute
SF-00280 Helsinki 28, FINLAND

Bacteroides buccae is among the few human oral species reported with an S-layer (Kornman and Holt 1981). The periodic arrangement of the lattice has been reported to be hexagonal (Ranta et al. 1983, Haapasalo et al. 1985, Sjögren et al. 1985).

Another human oral *Bacteroides* species, *B. heparinolyticus* with an outer layer was reported by Okuda et al. (1985), but no periodicity of this lattice was described. The aim of the present study was to compare the cell envelope ultrastructures of *B. heparinolyticus* and *B. buccae.*

Methods

Strains. For ultrastructural studies the *B. heparinolyticus* strain ATCC 35895^T was used. For comparison, two *B. buccae* strains were also studied, ATCC 33574, and strain ES57, which has been isolated from a necrotic root canal (Haapasalo et al. 1985). All strains were preserved in glycerin-milk at -70°C.

Electron microscopy. For thin sections the bacterial cells were collected from 3-day-old cultures from the MCG plates (Haapasalo et al. 1985) into 0.1 M phosphate buffer (pH 7.2) and the samples were prefixed with 3% glutaraldehyde and 4% tannic acid and prepared as previously described (Lounatmaa 1985).

For freeze-fracture and freeze-etching the cells were collected as described above and centrifuged (9980 *g,* 2 min, 20°C) to form a thick suspension and frozen in liquid Freon 22 cooled by liquid nitrogen. The fracturing in a Balzers freeze-etching unit BAF 400T (Balzers, Liechtenstein) was performed at -120°C and the platinum shadowing at the angle of 40° (Lounatmaa 1985).

The cells were negatively stained with 1% or 2% (w/v) phosphotungstic acid (pH 6.5) or 1% ammonium molybdate (pH 7.8). To produce fragments of S-layer, the cell suspension was mechanically pipetted 20 times (Kerosuo et al. 1987).

The electron micrographs were taken with JEM-100CX (negative staining) and with JEM-1200EX (thin sections and freeze-etched and freeze-fractured replicas) transmission electron microscopes at 80 kV and 60 kV, respectively.

Results and Discussion

In *B. heparinolyticus* the S-layer described as an outer layer by Okuda et al. (1985) seemed to consist of a narrow electron-dense layer surrounded by amorphous material. There was an approximately 20 nm gap between the outer membrane (OM) and the S-layer (Fig.1) as described also in *B. buccae* (Haapasalo et al. 1985), but in *B. heparinolyticus* this gap appeared more electron-dense.

A hexagonally oriented pattern with a centre-to-centre spacing approximately 20 nm was occasionally seen on the freeze-etched cells of *B. heparinolyticus* strain (Fig.2). In *B. buccae* strains the periodicity of the S-layer could only vaguely be seen, as reported earlier by Ranta et al. (1983).

In *B. buccae* the fracturing (without etching) revealed a periodic structure also in the outer leaflet of the OM, as described earlier (Haapasalo 1986, Kerosuo et al. 1987). This crystalline OM protein (COMP) was also hexagonally arranged and the centre-to-centre spacing was approximately 8 nm. In this study, the presence of COMP in *B. buccae* was frequently confirmed, but the same technique could not reveal any periodicity in the OM of *B. heparinolyticus* (Fig.3). Thus, the COMP found in *B. buccae* seems to be a unique finding so far.

The negatively stained cells of *B. heparinolyticus* showed delicate periodicity, although the type of the lattice could not be determined (Fig.4), whereas in *B. buccae* the hexagonal lattice was clearly seen, as reported previously (Sjögren et al. 1985).

By freeze-etching, the periodicity of the S-layer in a novel Gram-positive anaerobic rod could be clearly visualized (Kerosuo et al. 1988). The same technique could be applied in *B. heparinolyticus* (Fig.2), but in *B. buccae* the freeze-etching technique was less appropriate. This difference might be explained by the extracellular material, which has been shown to be even more abundant in many *B. buccae* strains (Haapasalo et al. 1985) compared to *B. heparinolyticus* (Fig.1).

Acknowledgement.
We thank Dr. Okuda for providing the *B. heparinolyticus* strain.

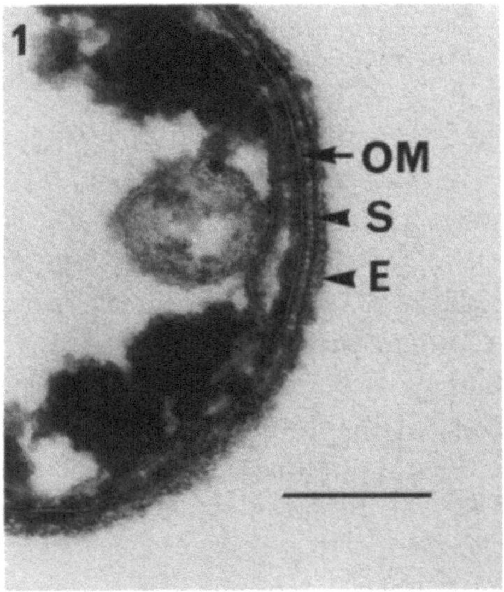

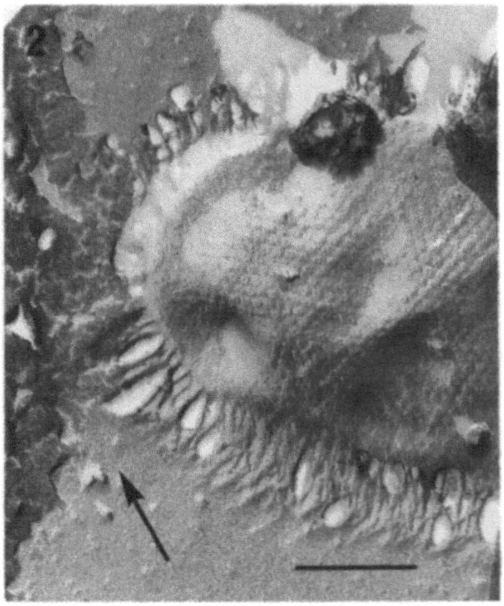

Fig.1. A thin sectioned cell of *B. heparinolyticus.* OM, outer membrane; S, surface-layer; E, extracellular material. Scale bar = 0.2 µm in this and subsequent micrographs.

Fig.2. A freeze-etched cell of *B. heparinolyticus.* A hexagonal periodicity with a centre-to-centre spacing of appr. 20 nm is seen. The direction of the shadowing is indicated by an arrow

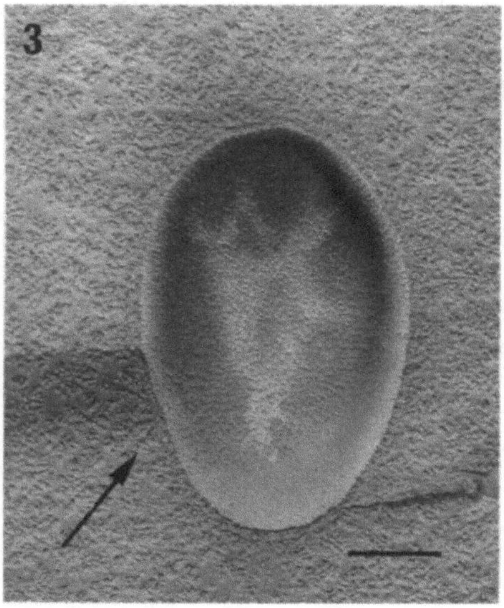

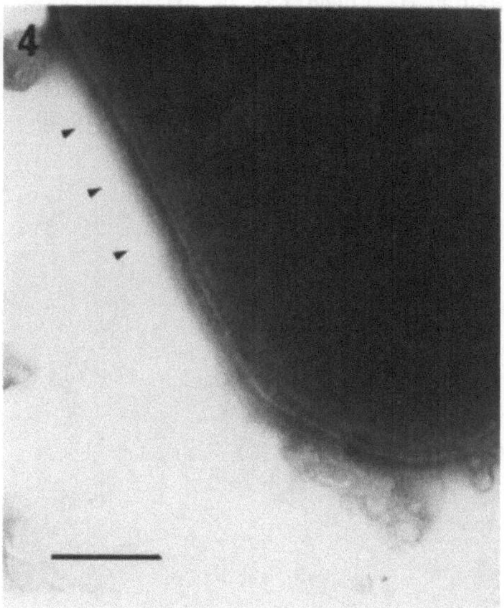

Fig.3. A freeze-fractured cell of *B. heparinolyticus.* Randomly distributed intramembranous particles are seen on the concave fracture face of the outer membrane. The direction of the shadowing is indicated by an arrow

Fig.4. A cell of *B. heparinolyticus* stained with 1% ammonium molybdate. Periodicity of the S-layer indicated by arrowheads

References

Haapasalo M (1986) *Bacteroides buccae* and related taxa in necrotic root canal infections. J Clin Microbiol 24:940-944

Haapasalo M, Lounatmaa K, Ranta H, Shah H, Ranta K (1985) Ultrastructure of *Bacteroides capillus, B. buccae, B. pentosaceus, B. oris, B. oralis, B. veroralis,* and pentose sugar-fermenting *Bacteroides* sp. from humans with periapical osteitis: occurrence of external proteinaceous cell wall layer. Int J Syst Bacteriol 35:65-72

Kerosuo E, Haapasalo M, Ranta H, Lounatmaa K (1987) Hexagonal periodicity in the outer membrane of *Bacteroides buccae.* J Gen Microbiol 133:2217-2224

Kerosuo E, Haapasalo M, Lounatmaa K, Ranta H, Ranta K (1988) Ultrastructure of a novel anaerobic Gram-positive nonsporing rod from dental root canal. Scand J Dent Res (in press)

Kornman KS, Holt SC (1981) Physiological and ultrastructural characterization of a new *Bacteroides* species *(Bacteroides capillus)* isolated from severe localized periodontitis. J Periodontal Res 16:542-555

Lounatmaa K (1985) Electron microscopic methods for the study of bacterial surface structures. In: Korhonen TK, Dawes EA, Mäkelä PH (eds) Enterobacterial surface antigens: methods for molecular characterization. Elsevier, Amsterdam, pp 243-261

Okuda K, Kato T, Shiozu J, Takazoe I, Nakamura T (1985) *Bacteroides heparinolyticus* sp. nov. isolated from humans with periodontitis. Int J Syst Bacteriol 35:438-442

Ranta H, Lounatmaa K, Haapasalo M, Ranta K (1983) Isolation and ultrastructure of *Bacteroides* sp. with external cell-wall layer (S-layer) in periapical osteitis. Scand J Dent Res 91:458-464
Sjögren A, Hovmöller S, Farrants G, Ranta H, Haapasalo M, Ranta K, Lounatmaa K (1985) Structures of two different surface layers found in six *Bacteroides* strains. J Bacteriol 164:1278-1282

S-Layer in Cyanobacteria

J. Šmarda

Department of Biology, Medical Faculty
J.E. Purkyně University
Obránců míru 10
CS-662 44 Brno
CZECHOSLOVAKIA

The first clear indication of a periodic structure of the cell wall surface layer in a Cyanobacterium was achieved by Schiewer and Jonas (1977) in transverse ultrathin sections of cells of a strain of *Synechocystis aquatilis.* Since then, eight reports have presented more detailed findings of S-layers in five different genera of Cyanobacteria. In our studies, we have so far investigated 15 species of 8 cyanobacterial genera. We applied the freeze-fracturing preparation techniques.

The objective of this chapter is to summarize briefly but completely all the reliable data available so far on the investigation of S-layers on Cyanobacteria, including our experience. It is reasonable to look for the S-layers only in unicellular species; all those are classified in a single order Chroococcales, with seven families (Komárek and Anagnostidis 1986).

Familiy: Microcystaceae

1. Subfamily: Aphanothecoideae

A periodic structure of the surface layer of one strain, *Aphanothece halophytica,* was reported (Simon 1981, according to Sleytr and Messner 1983); a lattice type was not distinguished.

Cells of the genus *Cyanothece* (one strain of each of the four species: *C. notata, C. eximia, C. minervae,* and *Cyanothece* sp.) were examined in ultrathin sections (Gromov et al. 1986). Only in *Cyanothece minervae* was an S-layer of square lattice observed. No regular surface structure was found in two strains of *Gloeothece* sp. (Vaara 1982).

2. Subfamily: Synechococcoideae

We were not able to detect any regular pattern on the cell wall surface of *Synechococcus elongatus, S. leopoliensis, S. vantieghemii* or

Synechococcus sp. (one strain of each was examined). Golecki (1979) could not distinguish any lattice on *S. lividus.* The only positive statement of Vaara (1982) that he found a regular hexagonal surface lattice on one of the 12 *Synechococcus* sp. strains examined, is probably due to an incorrect classification of this strain.

3. Subfamily: Merismopedioideae

We were not able to distinguish any paracrystalline surface lattice on: *Aphanocapsa grevillei, Aphanocapsa* sp., *Merismopedia glauca, M. punctata* (one strain of each was examined) or *Merismopedia* sp. (two strains examined).

However, considerable knowledge has been gathered on hexagonal lattice on *Synechocystis* cells. This was found first on one strain of *Synechocystis aquatilis* (Šmarda et al. 1979), being formed by uniform particles of some 10 nm in diameter, then by freeze-etching in a strain of *Synechocystis* sp. (Lounatmaa et al. 1980) with the lattice constant of 15.5 nm, and by freeze-fracturing in a further three of seven *Synechocystis* strains (Vaara 1982); finally we saw it also on one strain of *S. aquatilis* f. *salina* and on one strain of *S. fuscopigmentosa.* The dimensions and form of hexagonally arranged particles must not necessarily be the same in all cases. Karlsson et al. (1983) investigated the three-dimensional structure of the about 14-nm-thick S-layer on one strain of *Synechocystis* sp. in great detail: they found regular hexagonal arrays of 15.2 nm spacing. In each particle, six monomer protein molecules formed a hexamer arranged around a central hollow cylinder.

4. Subfamily: Microcystoideae

Jensen and Sicko noticed probably a regularly structured surface layer in a thin section of a strain of *Gloeocapsa alpicola* as early as 1972. We could see a regular pattern of identical cylindrical particles forming a surface layer on another strain of this species. From the genus *Microcystis,* a hexagonal surface pattern was noticed in ultrathin sections and freeze-fractured preparations of a strain of *M. firma* (Vaara 1982). On one of two strains of *M. incerta,* we proved a hexagonal lattice of a thick S-layer on only one of them. The column-like particles with distinct hexagonal capitals, thickened in the lower fifth of their height, were 8 nm in diameter and displayed a 11.3 nm spacing. The complete thickness of the layer of columns with capitals reached 46 nm. A hexagonal lattice was found in four strains of *M. marginata* (Kessel 1978 - spacing about 14 nm).

Table 1. Present conclusive evidence of the S-layer in cyanobacteria (survey)

Taxon	Lattice	Number of strains	Reference
Order: Chroococcales			
1 Family: Microcystaceae			
1.1 Subfamily: Aphanothecoideae			
Cyanothece minervae	p4	1	Gromov et al. (1986)
1.2 Subfamily: Synechococcoideae			
Synechococcus sp.[a]	p6	1	Vaara (1982)
1.3 Subfamily: Merismopedioideae			
Synechocystis aquatilis	p6	1	Šmarda et al. (1979)
Synechocystis aquatilis f. *salina*	p6	1	Šmarda (this chap.)
Synechocystis fuscopigmentosa	p6	1	Šmarda (this chap.)
Synechocystis sp.	p6	1	Lounatmaa et al. (1980)
Synechocystis sp.	p6	3	Vaara (1982)
1.4 Subfamily: Microcystoideae			
Microcystis firma	p6	1	Vaara (1982)
Microcystis incerta	p6	1	Šmarda (this chap.)
Microcystis marginata[a]	p6	4	Kessel (1978)
6 Family: Xenococcaceae			
Chroococcidiopsis sp.	p2	2	Büdel and Rhiel (1985)

[a]Presumably incorrect determination of the species.

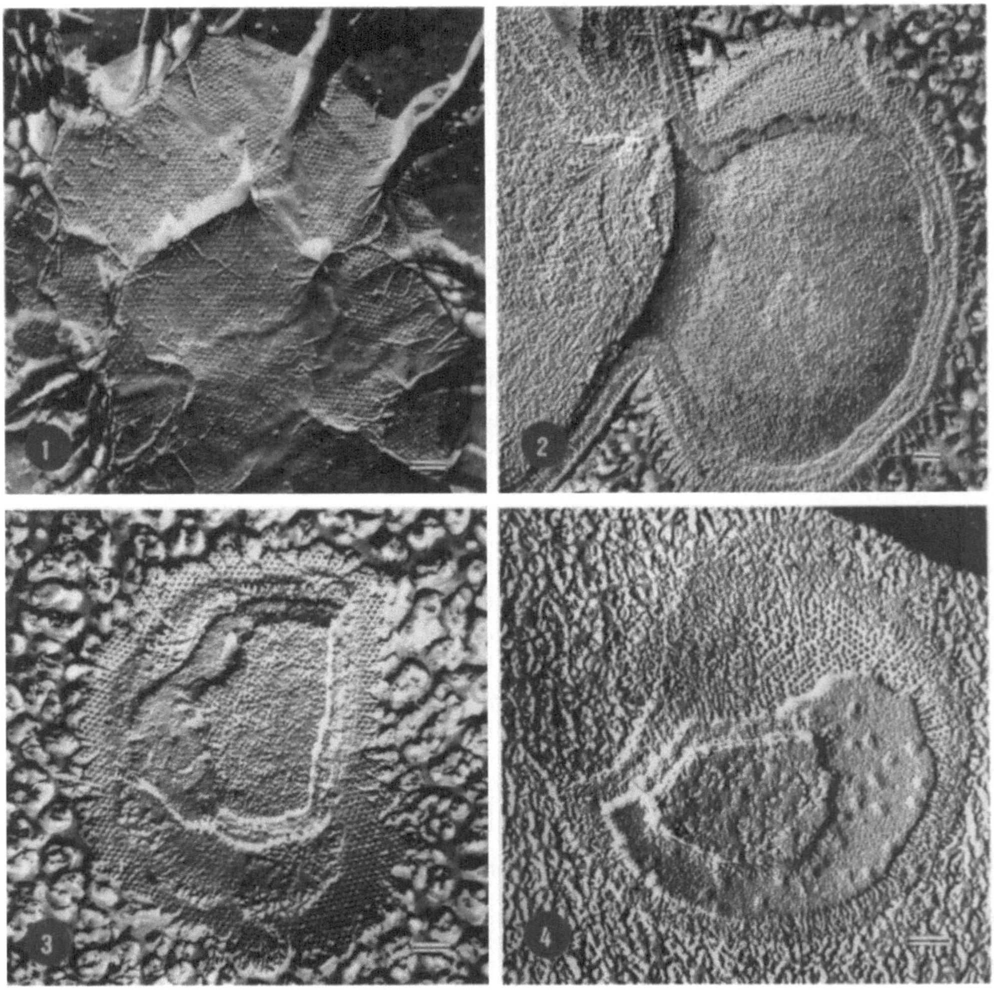

Figs.1–4. S-layer with a hexagonal lattice on cells of *Synechocystis aquatilis* strain Holubcová 1959/1. Bars = 100 nm

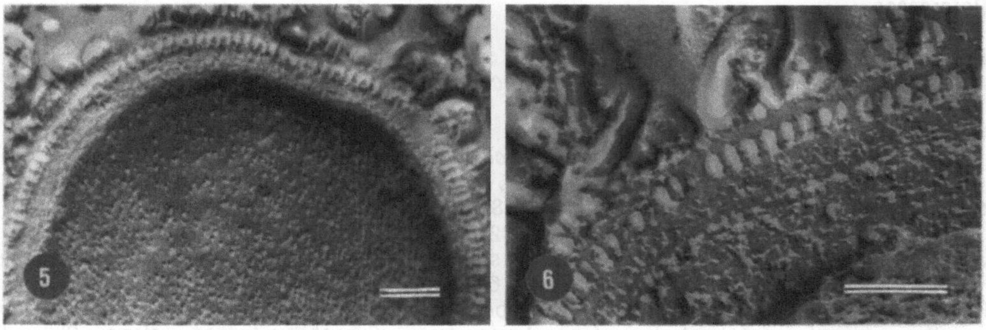

Figs.5-6. S-layer on cells of *Microcystis incerta* strain 151. Cell walls fractured perpendicularly. Bars = 100 nm

Family: Xenococcaceae

A peculiar pattern of the S-layer was found by freeze-fracturing in two strains of *Chroococcidiopsis* sp. (Büdel and Rhiel 1985). It was formed by uniform ribbon-like structures, 2-3 nm thick, striated at a right angle to their long axis. The parallel striation was due to rows of closely associated globular particles of 2.5-3 nm in diameter. The distance between the strips was 2-2.5 nm. We noticed no regular surface lattice on one strain of *Myxosarcina chroococcoides.*

Family: Chroococcaceae

In one strain of *Chroococcus minor,* we found no lattice of the wall surface.

All positive findings of hexagonal lattices concern subfamilies (3) and (4) of the family Microcystaceae, the finding of square lattice the subfamily (1) of it, and the finding of a quite different transverse row lattice the remote family Xenococcaceae. In any family or subfamily, only certain genera, certain species and merely certain strains seem to be able to bear S-layers.

References

Büdel B, Rhiel E (1985) A new cell wall structure in a symbiotic and a free-living strain of the blue-alga genus *Chroococcidiopsis* (Pleurocapsales). Arch Microbiol 143:117-121

Golecki JR (1979) Ultrastructure of cell walls and thylakoid membranes of the thermophilic cyanobacterium *Synechococcus lividus* under the influence of temperature shifts. Arch Microbiol 120:125-133

Gromov BV, Gavrilova OV, Konovalov ES (1986) Ultrastruktura kletok cianobakterij roda *Cyanothece*. Mikrobiologija 55:821-824

Jensen TE, Sicko LM (1972) The fine structure of the cell wall of *Gloeocapsa alpicola*, a blue-green alga. Cytobiology 6:439:446

Karlsson B, Vaara T, Lounatmaa K, Gyllenberg H (1983) Three-dimensional structure of the regularly constructed surface layer from *Synechocystis* sp. strain CLII. J Bacteriol 156:1338-1343

Kessel M (1978) A unique crystalline wall layer in the cyanobacterium *Microcystis marginata*. J Ultrastruct Res 62:203-212

Komárek J, Anagnostidis K (1986) Modern approach to the classification system of cyanophytes 2 - Chroococcales. Arch Hydrobiol Suppl 73(2):157-226

Lounatmaa K, Vaara T, Österlund K, Vaara M (1980) Ultrastructure of the cell wall of a *Synechocystis* strain. Can J Microbiol 26:204-208

Schiewer U, Jonas L (1977) Die Wirkung unterschiedlicher NaCl-Konzentrationen auf die Ultrastruktur von Blaualgen. II. *Synechocystis aquatilis*. Arch Protistenkd 119:146-162

Simon RD (1981) Gliding motility in *Aphanothece halophytica*: analysis of wall proteins in *mot* mutants. J Bacteriol 148:315-321

Sleytr UB, Messner P (1983) Crystalline surface layers on bacteria. Annu Rev Microbiol 37:311-339

Šmarda J, Čáslavská J, Komárek J (1979) Cell wall structure of *Synechocystis aquatilis* (Cyanophyceae). Arch Hydrobiol Suppl 56:154-165

Vaara T (1982) The outermost surface structures in chroococcacean cyanobacteria. Can J Microbiol 28:929-941

Cloning and Sequencing of the S-Layer Glycoprotein Gene of *Halobacterium halobium*

J. Lechner and M. Sumper

Biochemie I
Universität Regensburg
Universitätsstr. 31
D-8400 Regensburg
FRG

As early as 1956 Houwink demonstrated that the surface of halobacteria are covered by a hexagonally patterned macromolecular monolayer (surface layer, S-layer). Mescher and Strominger were the first authors to demonstrate the occurrence of a true glycoprotein as the main constituent of the S-layer in this prokaryotic organism (Mescher and Strominger 1976). Structural work following this discovery led to a rather detailed view of the saccharide structure and of the biosynthesis of this glycoprotein from *Halobacterium halobium* (Paul and Wieland 1987, Paul et al. 1986, Wieland et al. 1981, 1982, 1983, 1986, Lechner et al. 1985a,b). The cell surface glycoprotein (CSG) of halobacteria contains three types of protein-linked saccharides: (1) A single high molecular weight saccharide which is composed of repeats of sulphated pentasaccharides (Paul and Wieland 1987). It is linked to the polypeptide chain via the recently described new linkage Asn-GalNAc[1]. (2) About ten sulphated oligosaccharides (hexuronic acid 1 - $4)_{2-3}$Glc are connected to the polypeptide via Asn-Glc, another novel type N-glycosidic linkage (Wieland et al. 1983). (3) About 20 disaccharides Glc-Gal are O-glycosidically attached to threonine residues of the polypeptide chain (Mescher and Strominger 1976, Wieland et al. 1982).

The CSG-gene was cloned and sequenced as described in Lechner and Sumper (1987). Here we correlate the deduced CSG primary structure with the established structural data of the protein-linked saccharides.

In Fig.1 the results of all sequencing experiments are summarized. The 3167-bp segment shown was sequenced in both directions. An open reading frame (ORF) starts at position 457 with an ATG codon and ends at position 3015 with a TGA codon. The N-terminal amino acid sequence

TGATCGGTGGCGAACGGAGCCCCCGCATGGATGTTGTTACCCGCGCCGTCGCTCCCCGCGACG 63

GCCGGCTGTGCAGCGACACCGCCTGCAGGGACGACGAGCCCGGCGATGAGAACACAGACGCCG 126

ACGACGAGGCACTTCCGTGGCCCACTCCCGCTCATTGCGGCTTCGGTTCGAACTGGGGCATCT 189

TTAACCCCCGTTTTCGCCGGACGGCACCGTGCGTGATGTCGCCGCTCCGCCATCCACGTTTCA 252

TGTGAGCAATACACACCCAAATCGGTGTTCTGACGGCGGCTGACGCGCGAAAAGGCAGAAAGC 315

ATTTACCAGTGGCCGGGTATAGTCTGGAGCACCCCTACCCGAAATGGCGGCTGCAGAAACCCA 378

CGATTCACCGTTTCGGCGGAATCAGGTGGATGCGGTCGTCTGGTTGGACTGACACCGTAGCTC 441

-34
AGTCACTCAGTAAAA ATG ACA GAC ACA ACA GGC AAA CTC CGC GCA GTC CTC 492
Met Thr Asp Thr Thr Gly Lys Leu Arg Ala Val Leu

CTG ACG GCG CTG ATG GTC GGT TCC GTA ATC GGA GCC GGC GTC GCG TTC 540
Leu Thr Ala Leu Met Val Gly Ser Val Ile Gly Ala Gly Val Ala Phe

-1 1
ACG GGC GGG GCT GCT GCG GCG AAT GCA AGC GAC CTG AAC GAT TAT CAG 588
Thr Gly Gly Ala Ala Ala Ala Asn Ala Ser Asp Leu Asn Asp Tyr Gln

CGG TTC AAC GAA AAT ACA AAC TAC ACG TAT AGT ACC GCC TCA GAA GAC 636
Arg Phe Asn Glu Asn Thr Asn Tyr Thr Tyr Ser Thr Ala Ser Glu Asp

GGT AAA ACC GAA GGA AGT GTC GCC AGT GGC GCG ACC ATC TTC CAG GGC 684
Gly Lys Thr Glu Gly Ser Val Ala Ser Gly Ala Thr Ile Phe Gln Gly

50
GAA GAG GAC GTT ACC TTC CGG AAG CTG GAC AAC GAG AAA GAG GTC AGT 732
Glu Glu Asp Val Thr Phe Arg Lys Leu Asp Asn Glu Lys Glu Val Ser

CCG GCG ACC CTC TCC CGC ACT GGC GGG TCT GAC GAG GGC GTT CCT CTC 780
Pro Ala Thr Leu Ser Arg Thr Gly Gly Ser Asp Glu Gly Val Pro Leu

CAG ATG CCG ATC CCC GAG GAC CAG TCG ACC GGT TCC TAC GAT AGC AAT 828
Gln Met Pro Ile Pro Glu Asp Gln Ser Thr Gly Ser Tyr Asp Ser Asn

100
GGT CCA GAC AAC GAC GAG GCT GAC TTC GGC GTT ACG GTC CAG AGT CCA 876
Gly Pro Asp Asn Asp Glu Ala Asp Phe Gly Val Thr Val Gln Ser Pro

TCG GTG ACG ATG CTC GAA GTC CGC AAC AAC GCG GAC AAC GAC GTC ACC 924
Ser Val Thr Met Leu Glu Val Arg Asn Asn Ala Asp Asn Asp Val Thr

GGC GGT GTC CTG AAC ACA CAG CAG GAC GAG TTC TCG ATC GCC GTT GAC 972
Gly Gly Val Leu Asn Thr Gln Gln Asp Glu Ser Ser Ile Ala Val Asp

150
TAC AAC TAC TAC GCT GCC GAG GAC CTC GAG CTG ACC GTC GAA GAC GAG 1020
Tyr Asn Tyr Tyr Ala Ala Glu Asp Leu Glu Leu Thr Val Glu Asp Glu

GAC GGT CTC GAC GTT ACG GAC GAG ATC CTC GCT GCC GAC CAG TCG GGC 1068
Asp Gly Leu Asp Val Thr Asp Glu Ile Leu Ala Ala Asp Gln Ser Gly

GGC GCG TAC GAA GAT GGC ACC GGA AAC AAC GGG CCC AAC ACG CTT CGC 1116
Gly Ala Tyr Glu Asp Gly Thr Gly Asn Asn Gly Pro Asn Thr Leu Arg

200
TTC GAC ATC GAC CCG AAC AAC GTT GAC GCG GGC GAC TAC ACG GTC TCG 1164
Phe Asp Ile Asp Pro Asn Asn Val Asp Ala Gly Asp Tyr Thr Val Ser

GTT GAA GGT GTC GAG GAC CTG GAC TTC GGT GAC GCC ACC GAG TCC GCC 1212
Val Glu Gly Val Glu Asp Leu Asp Phe Gly Asp Ala Thr Glu Ser Ala

TCC GTG ACG ATT TCC TCC TCG AAC AAG GCA TCG CTG AAC CTC GCC GAG 1260
Ser Val Thr Ile Ser Ser Ser Asn Lys Ala Ser Leu Asn Leu Ala Glu
250
GAC GAA GTC GTG CAG GGA GCG AAC CTC AAG TAC ACC ATC GAG AAC AGT 1308
Asp Glu Val Val Gln Gly Ala Asn Leu Lys Tyr Thr Ile Glu Asn Ser

CCG GAA GGC AAC TAC CAC GCT GTC ACC ATC GAC AGC AGC GAC TTC CGC 1356
Pro Glu Gly Asn Tyr His Ala Val Thr Ile Asp Ser Ser Asp Phe Arg

GAC AGC AGC AGC GGT GCT GAT GCC GCG AAA GTC ATG CGC AGC GTT GGT 1404
Asp Ser Ser Ser Gly Ala Asp Ala Ala Lys Val Met Arg Ser Val Gly

GAC ACT GTC GAC ACC GGT CTC GTC GTC GAC AAC GAC AGT ACC ACC GAA 1452
Asp Thr Val Asp Thr Gly Leu Val Val Asp Asn Asp Ser Thr Thr Glu
300
ATT GTA GAC GAC TAT GAA AAC ACC TCG ATC TCG GAC GTC GAC TAC GCG 1500
Ile Val Asp Asp Tyr Glu Asn Thr Ser Ile Ser Asp Val Asp Tyr Ala

TAC GCC ATC GTC GAG ATC GAC GAC GGA AAC GGC GTC GGG TCC ATC GAG 1548
Tyr Ala Ile Val Glu Ile Asp Asp Gly Asn Gly Val Gly Ser Ile Glu

ACG CAG TAC CTC GAT GAC TCC AGC GCC GAC ATC GAC CTC TAC CCC GCA 1596
Thr Gln Tyr Leu Asp Asp Ser Ser Ala Asp Ile Asp Leu Tyr Pro Ala
350
TCC GAC ACC GAA GAC GCC CCG GAT TAC GTC AAT AGC AAC GAA GAA CTC 1644
Ser Asp Thr Glu Asp Ala Pro Asp Tyr Val Asn Ser Asn Glu Glu Leu

ACA AAC GGC TCC GCC CTC GAC GGC GTC TCT ACC GAC GAC GAC ACT GAC 1692
Thr Asn Gly Ser Ala Leu Asp Gly Val Ser Thr Asp Asp Asp Thr Asp

TTC GAC GTC ACG CAG GGC GAC ATC ACC CTC GAC AAC CCG ACC GGC GCG 1740
Phe Asp Val Thr Gln Gly Asp Ile Thr Leu Asp Asn Pro Thr Gly Ala
400
TAC GTT GTC GGC TCG GAA GTC GAC ATC AAC GGG ACC GCC AAC GAG GGG 1788
Tyr Val Val Gly Ser Glu Val Asp Ile Asn Gly Thr Ala Asn Glu Gly

ACT GAC GAC GTC GTG CTG TAC GCT CGC GAC AAC AAC GAC TTC GAA CTC 1836
Thr Asp Asp Val Val Leu Tyr Ala Arg Asp Asn Asn Asp Phe Glu Leu

GTC ACC GTC GAC GGT GAG AAA TCG ATC GAA GTC GAC AGC GAC GAC ACG 1884
Val Thr Val Asp Gly Glu Lys Ser Ile Glu Val Asp Ser Asp Asp Thr
450
TTC GAG GAG GAA GAC ATC ACG CTC TCC GAT GGA GAC AAG GGC GGT GAC 1932
Phe Glu Glu Glu Asp Ile Thr Leu Ser Asp Gly Asp Lys Gly Gly Asp

GAC ATC CTT GGT CTC CCC GGT ACG TAC CGC CTC GGC ATC ATC GCC AAG 1980
Asp Ile Leu Gly Leu Pro Gly Thr Tyr Arg Leu Gly Ile Ile Ala Lys

AGC GAC GCC GTC AAC TCG TCC GGT GGC GTG AAA GAC AAC ATT GAC ACG 2028
Ser Asp Ala Val Asn Ser Ser Gly Gly Val Lys Asp Asn Ile Asp Thr
500
TCC GAC TTC AAC CAG GGC GTC AGC AGT ACG TCC TCC ATC CGT GTG ACC 2076
Ser Asp Phe Asn Gln Gly Val Ser Ser Thr Ser Ser Ile Arg Val Thr

GAC ACG GAA CTC ACC GCG TCC TTC GAG ACC TAC AAC GGG CAG GTC GCC 2124
Asp Thr Glu Leu Thr Ala Ser Phe Glu Thr Tyr Asn Gly Gln Val Ala

```
GAC GAC GAC AAC CAG ATC GAC GTT GAG GGG ACT GCC CCT GGG AAG GAC 2172
Asp Asp Asp Asn Gln Ile Asp Val Glu Gly Thr Ala Pro Gly Lys Asp
                                            550
AAC GTT GCC GCC ATC ATC ATC GGC AGC CGT GGC AAG GTC AAG TTC CAG 2220
Asn Val Ala Ala Ile Ile Ile Gly Ser Arg Gly Lys Val Lys Phe Gln

TCC ATC TCC GTC GAC AGC GAC GAC ACG TTC GAC GAG GAG GAC ATC GAC 2268
Ser Ile Ser Val Asp Ser Asp Asp Thr Phe Asp Glu Glu Asp Ile Asp

ATC TCG GAG CTC CGA CAG GGC AGT GCT TCC GCA CAC ATC CTC TCG TCG 2316
Ile Ser Glu Leu Arg Gln Gly Ser Ala Ser Ala His Ile Leu Ser Ser
                                                    600
GGT CGT GAC GGG AAG TTC GGT GAG GAC ACC GCC AAC AGC ATT AGC GAT 2364
Gly Arg Asp Gly Lys Phe Gly Glu Asp Thr Ala Asn Ser Ile Ser Asp

CTT GAG GAC GAA GTC GGT AAC TAC ACC TCG GGT CCG CCG ACT GGC GAC 2412
Leu Glu Asp Glu Val Gly Asn Tyr Thr Ser Gly Pro Pro Thr Gly Asp

CAG ATC CGC GAC CGC ATC CTC TCG AAC ACG GTC GAC GAC ACC GCC AGC 2460
Gln Ile Arg Asp Arg Ile Leu Ser Asn Thr Val Asp Asp Thr Ala Ser
                                                        650
GAC GAC CTC ATC GTC ACG CAG CAG TTC CGT CTC GTT GAC GGA CTC ACC 2508
Asp Asp Leu Ile Val Thr Gln Gln Phe Arg Leu Val Asp Gly Leu Thr

ACG ATC GAA GCC ACT GAG GGT GGC GAA GCG GGC GGC TCG GTC ACC GTC 2556
Thr Ile Glu Ala Thr Glu Gly Gly Glu Ala Gly Gly Ser Val Thr Val

ATG GGG ACG ACC AAC CGG AAG GCC GAC GAC AAC ACC ATC ACG GTT GAA 2604
Met Gly Thr Thr Asn Arg Lys Ala Asp Asp Asn Thr Ile Thr Val Glu

CTC CTC CAG GGC GAC GCG TCC ATC GAG ATC AAC AGC ACT GAT GAG TGG 2652
Leu Leu Gln Gly Asp Ala Ser Ile Glu Ile Asn Ser Thr Asp Glu Trp
    700
AAC AGC GAC GGC CAG TGG TCG GTT GAT GTC CCG CTC TCG AAC GTC GAG 2700
Asn Ser Asp Gly Gln Trp Ser Val Asp Val Pro Leu Ser Asn Val Glu

CCG GGC AAC TAC ACG GTC GAA GCT GAC GAC GGT GAC AAC ACC GAC CGT 2748
Pro Gly Asn Tyr Thr Val Glu Ala Asp Asp Gly Asp Asn Thr Asp Arg

CAG AAC GTC GAA ATC GTC GAG GAA CTC GAG GAG CCT GAT CAG ACG ACC 2796
Gln Asn Val Glu Ile Val Glu Glu Leu Glu Glu Pro Asp Gln Thr Thr

            750
GTC GAT CAG CCC GAG AAC AAC CAG ACG ATG ACG ACG ACG ATG ACC GAG 2844
Val Asp Gln Pro Glu Asn Asn Gln Thr Met Thr Thr Thr Met Thr Glu

ACG ACC ACC GAG ACG ACC ACC GAG ATG ACC ACC ACG CAG GAG AAC ACC 2892
Thr Thr Thr Glu Thr Thr Thr Glu Met Thr Thr Thr Gln Glu Asn Thr

ACC GAG AAC GGC TCG GAG GGC ACT TCC GAT GGC GAG TCA GGC GGC AGC 2940
Thr Glu Asn Gly Ser Glu Gly Thr Ser Asp Gly Glu Ser Gly Gly Ser
                    800
ATC CCC GGC TTC GGT GTC GGT GTT GCG CTC GTC GCG GTC CTC GGT GCG 2988
Ile Pro Gly Phe Gly Val Gly Val Ala Leu Val Ala Val Leu Gly Ala

GCG CTG CTG GCA CTC CGC CAG AAC TGA TTGACCCACTGAATCACGTCTGACCGC 3042
Ala Leu Leu Ala Leu Arg Gln Asn

GCGTACGCGGTCACTTGCGGTGCCGTTTTCTTTGTTACCGACGACCGACCAGCGACAGCCACC 3105

GCGCGCTCACTGCCACCAAAAGAGTCATATCACAGCCGACCAGTTTCTGGAACGTTCCCGAT 3167
```

Fig.1. DNA sequence of CSG gene. The DNA sequence was determined in both directions. The numbers on the right indicate nucleotide positions. Numbers above the nucleotide sequence designate amino acid positions. Numbers above the nucleotide sequence designate amino acid positions. The N-terminal amino acid of the mature protein was marked "1". Negative values indicate a leader sequence. Sequences which were confirmed by peptide sequencing are underlined. Boxed regions represent potential N-glycosylation sites and arrows indicate self-complementary nucleotide sequences

of CSG was determined earlier and found to be: Ala-Asn-Ala-Ser-Asp-Leu-Asn-Asp-Tyr-Gln-Arg-Phe-Asn-Glu-Asn-Thr-Asn-Tyr-Thr-Tyr-Ser- (Paul et al. 1986). This sequence starts with the 35th amino acid of the ORF (position 1 in Fig.1). Three other stretches of the predicted amino acid sequence (underlined in Fig.1) were confirmed by sequencing of pure peptides. This allows the firm conclusion that the translation of the ORF indeed represents the primary structure of CSG. The sequence downstream of the translational stop codon TGA is typical for terminator structures. A GC-rich self-complementary region (stem and loop, see arrows in Fig.2) is followed by an AT-rich sequence. The ORF starts 102 bp upstream of the N-terminus of mature CSG, indicating the occurrence of a signal sequence on the nascent translation product. The open reading frame upstream of the N-terminus includes two methionines: Met(-34) and Met(-18). Therefore, the initiation site of CSG mRNA synthesis was determined by a primer extension experiment and found to be the adenosine nucleotide at bp 346. Assuming that the first methionine encoded by the mRNA is used to start translation, the protein chain is initiated with methionine (-34). Thus, the CSG gene encodes for a leader peptide being absent in the mature glycoprotein.

Twelve potential N-glycosylation sites (sequons Asn-X-Ser(Thr); X≠ Asp, Pro) are distributed throughout the polypeptide chain. The asparagine at position 2 is linked to the repetitive pentasaccharide structure occurring once per glycoprotein molecule (Paul et al. 1986). From the chemical data obtained previously it was concluded that about ten sulphated saccharides of the type (hexuronic acid 1-4)$_{2-3}$Glc are linked N-glycosidically to the polypeptide chain (Wieland et al. 1983). These data imply that most or all potential glydcosylation sites are indeed linked to saccharides. Glycopeptide squence analysis has directly confirmed glycosylation at Asn 2, Asn 479 and Asn 609. However, from N-terminal protein sequencing it was found that the sequon at position Asn 17 does not serve as a saccharide acceptor. It was found previously that all of the disaccharides linked O-glycosidically to CSG via threonine can be recovered in a single glycopeptide after pronase digestion. This indicated a highly clustered arrangement of the O-linked disaccharides. The DNA sequencing data confirmed this arrangement: 14 Thr residues are clustered within a stretch of 20 amino acids. Since CSG carries about 20 disaccharide units per molecule, all of these Thr residues within this cluster are likely to be engaged in O-glycosidic linkages. The entire polypeptide chain of the mature glycoprotein shows a single highly hydrophobic stretch of 21 amino acids (positions 795 to 815) which is

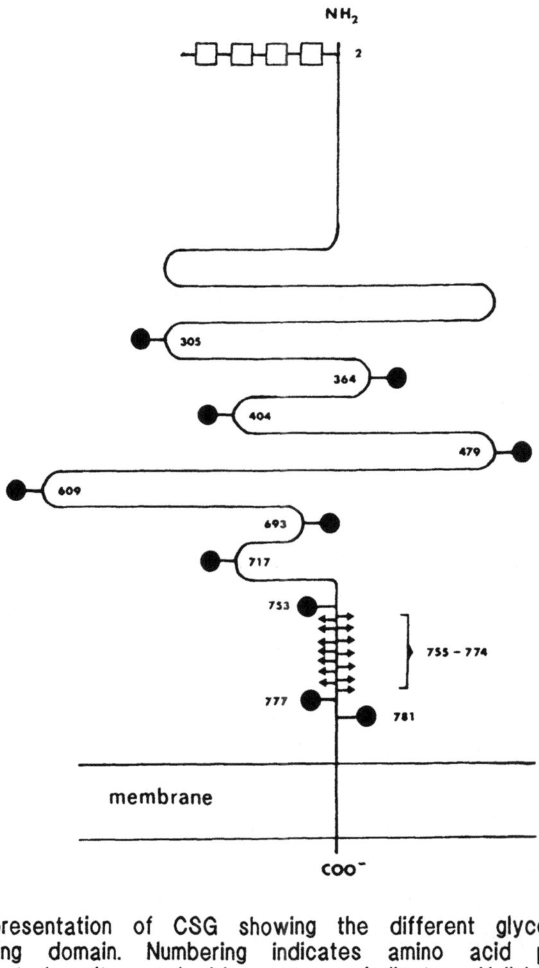

Fig.2. Schematic representation of CSG showing the different glycosylation sites and the membrane binding domain. Numbering indicates amino acid positions. —□—□— represents the repeated unit saccharide. ●— indicates N-linked oligosaccharides (hexuronic acid 1-4)$_{2-3}$Glc. ← indicates O-linked Glc-Gal disaccharides. Amino acids at position 2, 479, 609, 755, 757, 758, 759, and 761 were confirmed to be glycosylated by sequencing of the corresponding glycopeptides. For further explanations see text

only three amino acid positions away from the C-terminus. Most probably, this hydrophobic peptide serves as a membrane anchor. All other regions of the polypeptide chain of the mature CSG mainly consist of polar amino acid residues and show a dominant negative net charge. From the predicted amino acid sequence for the mature CSG protein a molecular mass of 86,538 daltons is calculated. Figure 2 schematically summarizes the structural features of CSG with respect to saccharide attachment sites and the hydrophobic peptide stretch.

References

Houwink AL (1956) Flagella, gas vacuoles and cell-wall structure in *Halobacterium halobium;* an electron microscope study. J Gen Microbiol 15:146-150

Lechner J, Sumper M (1987) The primary structure of a procaryotic glycoprotein. Cloning and sequencing of the cell surface glycoprotein gene of halobacteria. J Biol Chem 262:9724-9729

Lechner J, Wieland F., Sumper M (1985a) Biosynthesis of sulfated saccharides N-glycosidically linked to the protein via glucose. Purification and identification of sulfated dolichyl monophosphoryl tetrasaccharides from halobacteria. J Biol Chem 260:860-866

Lechner J, Wieland F, Sumper M (1985b) Transient methylation of dolichyl oligosaccharides is an obligatory step in halobacterial sulfated glycoprotein biosynthesis. J Biol Chem 260:8984-8989

Mescher MF, Strominger JL (1976) Purification and characterization of a prokaryotic glycoprotein from the cell envelope of *Halobacterium salinarium.* J Biol Chem 251:2005-2014

Paul G, Wieland F (1987) Sequence of the halobacterial glycosaminoglycan. J Biol Chem 262:9587-9593

Paul G, Lottspeich F, Wieland F (1986) Asparaginyl-N-acetylgalactosamine. J Biol Chem 261:1020-1024

Wieland F, Lechner J, Bernhardt G, Sumper M (1981) Sulphation of a repetitive saccharide in halobacterial cell wall glycoprotein. Occurrence of a sulphated lipid-linked precursor. FEBS Lett 132:319-323

Wieland F, Lechner J, Bernhardt G, Sumper M (1982) The cell wall glycoprotein of halobacteria: structural, functional and biosynthetic aspects. Zentralbl Bakteriol Mikrobiol Hyg 1 Abt Orig C3:161-170

Wieland F, Heitzer R, Schaefer W (1983) Asparaginylglucose: novel type of carbohydrate linkage. Proc Natl Acad Sci USA 80:5470-5474

Wieland F, Lechner J, Sumper M (1986) Iduronic acid: constituent of sulphated dolichyl phosphate oligosaccharides in halobacteria. FEBS Lett 195:77-81

The Primary Structure of the HPI-Layer Polypeptide of *Deinococcus radiodurans*

J. Peters, M. Peters, F. Lottspeich, W. Schäfer, Z. Cejka and W. Baumeister

Max-Planck-Institut für Biochemie
D-8033 Martinsried bei München
FRG

Deinococcus radiodurans possesses a complex cell envelope which is quite unique regarding both structure and composition (Thompson et al. 1982, Brooks et al. 1980). The structure determination of its S-layer (HPI-layer) is particularly advanced (Baumeister et al. 1986, Rachel et al. 1986). As part of our objective to obtain a high-resultion 3-D structure of the HPI-layer we have determined the primary structure of the HPI-layer polypeptide.

The gene coding for the HPI-polypeptide of *Deinococcus radiodurans* was cloned in *E. coli* using the plasmid pUC18, expressed under control of a *Deinococcus* promotor and detected by specific polyclonal antibodies on Western blots (see Peters and Baumeister 1986). The complete nucleotide sequence of the gene was determined. About one third of the sequence was confirmed by N-terminal amino acid sequence analysis (Peters et al. 1987). The nascent polypeptide comprises 1036 amino acids (Fig.1). We do not know the precise location of the leader peptidase cleavage site because the N-terminus is blocked to Edman degradation. However, from deletion studies of the gene and from amino acid sequence data we may infer that there is a leader sequence which is cleaved between position 20 and 59 in *E. coli.* From the patterns of proteolytic fragmentation of the HPI-polypeptide in situ and from studies with spheroplasts we conclude that the complete polypeptide is exported through the cytoplasmic membrane in *E. coli.* Polypeptides with truncated leader sequences are produced as *lac z'* fusion polypeptides.

The amino acid sequence of the HPI-layer polypeptide has some salient features: (1) It contains six cysteine residues. The disulphide bridges Cys-74/Cys-86 and Cys-642/Cys-754 were identified.

10 30 50

MKKNIALMALTGVLTLASCGQNGNTPTADTTAPTVSLSVNNANLPSGVGSVVLSGTVNEA

70 90 110

STVVVKNNAGTTVCTVEVAASGTFTCPATTIAGNTSTTSTSTSYTATATDAAKNVGTSSV

130 150 170

VTVNVAGVSNPAPTTAVLTIDLAGVSSAPITIKDANGNVVQGYDNVTVNDNATITVARGV

190 210 230

YTVTAGNVSGFNGPTTNFRVDLSGGNQTVTLNYTQAGTTTPTPVGSINILTPAVGTSVTG

250 270 290

GSTVRVTFDKANEVQCMVGGAAAVTAQVDSTSGYCDVVVPNSTGNVVITVMGKGVNGQTV

310 330 350

TATRNISVTQAAVSYGVVTPAGDQELTLTSEGIVRDADSGWRRLGQGVSTPSDPNLNLDI

370 390 410

YIKGTVNFSVNAPAGQKVELFLARTTGSDVPTNDDIQAGDVLRSVASTSGTETFSLDSRR

430 450 470

LAEFDGVRKWIVVRINGTQVTYQPVIADNKGPQQPDPELNGVQNAYSNILNNYNNSGLTY

490 510 530

VRGPVNVFTSNPSLQDREFGQAPVGSSFVQRRPAGFESIRYYLVPESAFNNKALQESDEM

550 570 590

LRAKAVKSVATVVSAPVLEPGTVKATSFSRVIGSGATSTVAPKALDNVTYRVYAISRDQV

610 630 650

GNETASATYDLVRFDNVGPTITGSVIRDTSDLPFPSQEPERCLSDIATISLGGIADNVGG

670 690 710

VGLNPGQGLTFTLGGRQIQAGQFDTNQLADPEYTIGFNSLTDALGNPVVTAPTNAKVYID

730 750 770

NTDPTVNFNRAVMQGTYASGGRVSVESDASDGGCGVYETRLFWDTANGVVDDATTTPAIG

790 810 830

HPVQFARQRVTDGAKADSLNAGWNALQLPNGAGAVYLRSLVVDRAGNATISTTSIVVNAK

850 870 890

ITNQARPLLGGFDAFKRNASAQFVGDDNVIAGVNGTAATPNVTGNSALDNILSLDSVGTL

910 930 950

TTNAYLPRGATETAITEKIRNVGAYGRFDATQWNLIRDYQLNTDPTLRSAYVNAGNLANQ

970 990 1010

RGNNWRIRTPWVELGSSDTANTQQKFDFNSDLLNDFYYGRTFGNNHSVNLFSYDQFNGVV

1030

SDTAGAYSFYGETVRK

Fig.1. Amino acid sequence of the nascent HPI-layer polypeptide as derived from the nucleotide sequence and 35% of the amino acid sequence

(2) The content of the hydroxyamino acids serine and threonine is very high (19.4%), particularly in the N-terminal region up to position 250 (29%). This region also contains a cluster of nine consecutive Ser/Thr residues which is unique for bacterial proteins thus far available for computer analysis. (3) The C-terminal region is extremely rich in aromatic amino acid residues.

The major part of the Asx and Glx is present as Asn and Gln. Therefore the net negative charge is only 16 per molecule. The overall distribution of hydrophobic and charged residues compares to that of typical outer membrane proteins. Thus, there are no hydrophobic core regions characteristic of cytoplasmic membrane proteins. The chemical characterization of the HPI-layer revealed that the protein contains both tightly bound reducing sugars (approx. 5 glucose and 1 galactose residue and traces of mannose) and fatty acids (1.4 per polypeptide chain, predominantly 16:1 and also 16:0, 15:1, 15:0, 17:1, and 17:0). The fatty acid is located in the N-terminal region.

The secondary structure predictions for the HPI-polypeptide yielded very good overall agreement with the experimental data derived from infrared spectroscopic measurements which indicated about 30% β-structure but virtually no α-helix content.

In the search for distant evolutionary relationships we compared the amino acid sequence of the HPI-polypeptide to that of the outer wall polypeptide of *Bacillus brevis* (Tsuboi et al. 1986) and the S-layer of *Halobacterium halobium* (Lechner and Sumper 1987). We found a statistically significant local homology between two segments of the HPI-polypeptide and the outer wall polypeptide (Fig.2)

```
      814                                                                      889
HPI   AVYLRSLVVDRAGNATISTTSIVVNAKITNQARPLLGGFDAFKRNASAQFVGDDNVIAGVNGTAATPNVTGNSALD
        •    •       ••  •                 •    •••••  •                 •
      • •  • • •••• •••••••• •••• •••• •••••••••••••••••  •••• •  • • •••••• ••• •
        •    •       ••  •                 •    •••••  •                 •
OWP   EGYTIAGVETKGEVVTIKVTELDIDDSDATPTVAVIGSVEDLKRNASGPFQPQKAIDGVSAPDKEAPVTGVEAGKT
      767                                                                      843
```

Fig.2. Sequence comparison (see Barker and Dayhoff 1972) of segments from the HPI-layer polypeptide (HPI) and the outer wall polypeptide (OWP) (Tsuboi et al. 1986). Triple dots indicate identical matches, single dots mark conservative replacements. The S.D. value of 6.5 corresponds to a statistically significant homology

One interesting aspect of S-layer structure is the interaction between the S-layer and the outer membrane. From the data of Thompson et al. (1982) we know that the HPI-layer is associated with the underlying outer membrane via hydrophobic forces. We found that the association of the outer membrane bound 5'-3'-exonuclease, the most abundant protein of the cell envelope, with the HPI-layer should contribute substantially to these hydrophobic interactions. By titration of the outer membrane with SDS at $0^{o}C$ and subsequent Percoll density gradient centrifugation, we obtained a red pigmented band which contained the HPI-polypeptide and the exonuclease in a 1:1 (+20%) stoichiometry. The topographical relationship between the two polypeptides was also established by reversible chemical cross-linking with 3,3'-dithiobis(succinimidyl propionate). Three-dimensional reconstruction of negatively stained complete envelopes revealed that the surface of the HPI-layer facing the outer membrane is covered with hexamers of particles with a M_r of at least 80,000 each. We are now attempting to perform image reconstructions of the delipidated HPI: exonuclease protein complex. When the purified HPI-layer was digested with trypsin, the N-terminal region of the HPI-polypeptide was cleaved at Lys-113, Lys-153, Arg-178, and Arg-199. However, when complete envelopes were treated with trypsin, these sites were inaccessible to the protease, indicating that the N-terminal region of the HPI-polypeptide interacts with the outer membrane and/or the membrane-bound exonuclease. As this region of the polypeptide contains the fatty acid, it might serve as an additional membrane anchor.

Further studies on the HPI-layer protein will include epitope mapping with antibodies directed against synthetic peptides and attempts to express derivatives of the cloned HPI-gene in a HPI-deficient mutant strain of *Deinococcus radiodurans.* Directed mutagenesis should be useful in addressing questions relating to the assembly of the HPI-layer.

References

Barker WC, Dayhoff MO (1972) Detecting distant relationships: computer methods and results. In: Dayhoff MO (ed) Atlas of protein sequence and structure. Natl Biomed Res Found, Washington DC, pp 101-110

Baumeister W, Barth M, Hegerl R, Guckenberger R, Hahn M, Saxton OW (1986) The three-dimensional structure of the regular surface layer (HPI-layer) of *Deinococcus radiodurans.* J Mol Biol 187:241-253

Brooks BW, Murray RGE, Johnson JL, Stackebrandt E, Woese CR, Fox GE (1980) Red-pigmented micrococci: a basis for taxonomy. Int J Syst Bacteriol 30:627-646

Lechner J, Sumper M (1987) The primary structure of a procaryotic glycoprotein. J Biol Chem 262:9424-9729
Peters J, Baumeister W (1986) Molecular cloning, expression, and characterization of the gene for the surface (HPI-) layer protein of *Deinococcus radiodurans* in *Escherichia coli.* J Bacteriol 167:1048-1054
Peters J, Peters M, Lottspeich F, Schaefer W, Baumeister W (1987) Nucleotide sequence analysis of the gene encoding the *Deinococcus radiodurans* surface protein, derived amino acid sequence, and complementary protein chemical studies. J Bacteriol 169:5216-5223
Rachel R, Jakubowski U, Tietz H, Hegerl R, Baumeister W (1986) Projected structure of the surface protein of *Deinococcus radiodurans* determined to 8 Å resolution by cryoelectronmicroscopy. Ultramicroscopy 20:305-316
Thompson BG, Murray RGE, Boyce JF (1982) The association of the surface array and the outer membrane of *Deinococcus radiodurans.* Can J Microbiol 28:1081-1088
Tsuboi A, Uchihi R, Tabata R, Takahashi Y, Hashiba H, Sasaki T, Yamagata H, Tsukagoshi N, Udaka S (1986) Characterization of the genes coding for two major cell wall proteins from protein-protecting *Bacillus brevis* 47. Complete nucleotide sequence of the outer wall protein gene. J Bacteriol 168:365-373

Characterization and Application of S-Layer Protein Gene for Production of Foreign Proteins in a Protein-Producing *Bacillus brevis* 47

N. Tsukagoshi

Department of Food Science and Technology
Faculty of Agriculture, Nagoya University
Nagoya 464
JAPAN

The synthesis and secretion of proteins by bacteria have long been of interest (Glenn 1976, Priest 1977). The process of secretion is selective and efficient. All cells must have mechanisms for sorting and distributing proteins to their correct locations. *Bacillus brevis* 47 seems to be a useful organism for studying the roles of the cell surface structure in protein secretion, since *B. brevis* 47 is the first organism that was found to produce vast amounts of protein, up to 12 g/l under optimal growth conditions (Udaka 1976).

General Properties of *B. brevis* 47

B. brevis 47 has a morphologically unique cell wall structure consisting of two protein layers and a peptidoglycan layer (Yamada et al. 1981). The outer two protein layers, termed the outer wall (OW) and middle wall (MW) layers, are composed of proteins with approximate molecular weights of 130 kDa and 150 kDa, respectively, as judged by SDS-PAGE. Each of the outer- and middle-wall proteins (OWP and MWP) forms a hexagonal array with a lattice constant of 14.5 nm in the cell wall (Fig.1) (Tsukagoshi et al. 1982, Tsuboi et al. 1982). The fine cell wall structure of *B. brevis* 47 is morphologically altered to various extents depending on the growth periods. Cells shed the outer two layers concomitantly with a prominent increase in protein secretion. The protein secreted consists mainly of two proteins with molecular weights of 130 and 150 kDa. Based on morphological, immunological, and biochemical criteria, the two major extracellular proteins were found to be derived from the respective cell wall proteins. There was, however,

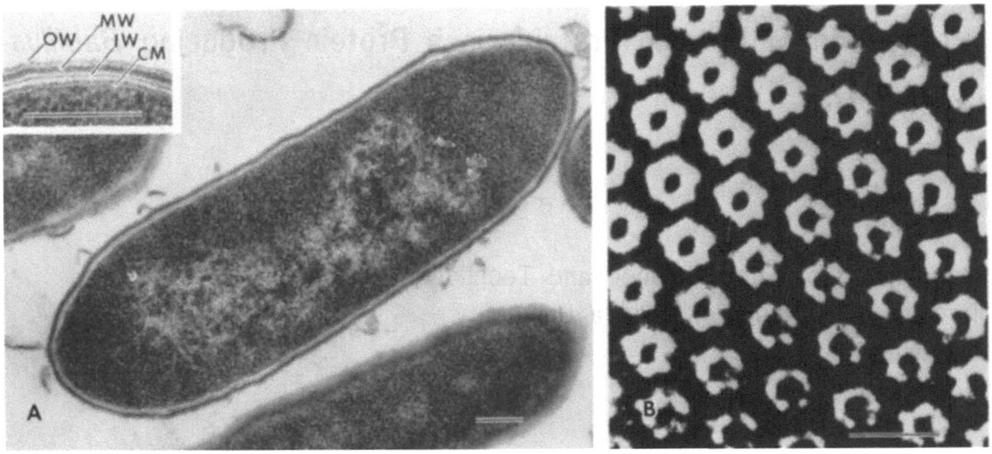

Fig.1. Cell wall structure (A) of *B. brevis* 47 and one-sided filtered image (B) of the cell envelopes. Bars in A and B = 200 and 20 nm, respectively

no serological relationship between these major proteins (Ohmizu et al. 1983).

Cloning and Characterization of Cell Wall Protein Genes in *B. brevis* 47

Immunological procedures were used to clone the genes coding for the cell wall proteins, since no biological activities were detected for either protein. Portions of the structural genes for the MWP and OWP were cloned in *E. coli* with pBR322 as a vector (pNT200) and found to direct the synthesis of polypeptides reactive with anti-MWP serum and ones reactive with anti-OWP serum (Tsukagoshi et al. 1984). With the cloned DNA fragment as a hybridization probe, the restriction map of the region surrounding the cell wall protein genes on the *B. brevis* 47 genome was constructed, and it showed that these genes form a gene cluster. Northern blot analysis of *B. brevis* 47 RNA indicated that these genes constitute a cotranscriptional unit and that they are transcribed from a promoter(s) located upstream of the MWP gene (Tsuboi et al. 1986, Yamagata et al. 1987). Therefore, the cell wall protein genes of *B. brevis* 47 seem to constitute an operon under coordinated control of their expression.

We determined the nucleotide sequence of both cell wall protein genes. MW and OW protein genes consisted of 3090 bp (1030 amino acid residues with a molecular weight of 114,830) and 2940 bp (980 amino acid residues with a molecular weight of 103,740), respectively. Both MWP and OWP were synthesized in precursor forms with signal peptides of 23 and 24 amino acids, respectively. When the hydrophathy profiles of both proteins were calculated based on the amino acid sequences deduced from the nucleotide sequences, MWP appears to be surprisingly hydrophilic compared to OWP except for the NH_2-terminal region. The charged amino acid residues are distributed rather densely and uniformly in MWP. We are in the process to determine which region(s) of the cell wall proteins is essential to form the paracrystalline structure by a combination of recombinant DNA and physical techniques.

Application of S-Layer Protein Gene for Foreign Protein Production in *B. brevis* 47

Since *B. brevis* 47 has the ability to produce extracellular proteins in amounts thus far not found for other microorganisms, we examined at first the efficiency of the production of a foreign protein, a thermophilic α-amylase in this organism (Tsukagoshi et al. 1985). *B. brevis* 47 carrying the *B. stearothermophilus* α-amylase gene on pUB110 (pBAM101) produced the enzyme most efficiently of the hosts examined, 100, 15 and 5 times more than *B. stearothermophilus, E. coli* HB101 (pHI301), and *B. subtilis* IA289 (pBAM101), respectively. The amount of the enzyme produced by *B. brevis* 47 (pBAM101) was approximately 0.5 g/l.

As the S-layer protein genes are expressed quite efficiently in *B. brevis* 47, we tried to utilize the 5′ region of the S-layer protein genes for foreign protein production. The translational fusion of the 5′ region of MWP gene with the *B. licheniformis* α-amylase gene resulted in the efficient expression of the α-amylase gene in *B. brevis* 47 (Yamagata et al. 1987). The amount of the enzyme produced was approximately 0.7 g/l. Furthermore, the *B. macerans* CGTase was produced at 1 g/l when the gene was expressed under the promoter of the cell wall protein gene operon.

Based on these results, pNU100, a secretion vector, was constructed by inserting the 600 bp *AluI-AluI* fragment containing both the promoter of the cell wall protein gene operon and the signal peptide sequence of the MWP gene onto the plasmid containing the erythromycin resistance gene and the replication origin of pUB110.

Currently, we are examining the productivity of eukaryotic proteins in *B. brevis* 47 utilizing pNU100 as a secretion vector. *B. brevis* 47 carrying swine pepsinogen cDNA on pNU100 produced extracellularly swine pepsinogen, of which the acid protease activity was completely inhibited by pepstatin, a specific pepsin inhibitor. The amount of the enzyme protein produced was rather small, 1 mg/l. However, *B. brevis* HPD31, a newly isolated protein-producing bacterium, transformed with the same plasmid produced approximately 10 mg/l of the enzyme. We are in the process of increasing the productivity of eukaryotic proteins in *B. brevis* 47 and related organisms.

References

Glenn AR (1976) Production of extracellular proteins by bacteria. Annu Rev Microbiol 30:41-62

Ohmizu H, Sasaki T, Tsukagoshi N, Udaka S, Kaneda N, Yagi K (1983) Major proteins released by a protein-producing bacterium, *Bacillus brevis* 47, are derived from cell wall protein. J Biochem 94:1077-1084

Priest FG (1977) Extracellular enzyme synthesis in the genes *Bacillus.* Bacteriol Rev 41:711-753

Tsuboi A, Tsukagoshi N, Udaka S (1982) Reassembly in vitro of hexagonal surface arrays in a protein-producing bacterium, *Bacillus brevis* 47. J Bacteriol 151:1485-1497

Tsuboi A, Uchihi R, Tabata R, Takahashi Y, Hashiba H, Sasaki T, Yamagata H, Tsukagoshi N, Udaka S (1986) Characterization of the genes coding for two major cell wall proteins from protein-producing *Bacillus brevis* 47: Complete nucleotide sequence of the outer wall protein gene. J Bacteriol 168:365-373

Tsukagoshi N, Yamada H, Tsuboi A, Udaka S, Katsura I (1982) Hexagonal surface array in a protein-secreting bacterium, *Bacillus brevis* 47. Biochim Biophys Acta 693:134-142

Tsukagoshi N, Tabata R, Takemura T, Yamagata H, Udaka S (1984) Molecular cloning of a major cell wall protein gene from protein-producing *Bacillus brevis* 47 and its expression in *E. coli* and *B. subtilis.* J Bacteriol 158:1054-1060

Tsukagoshi N, Iritani S, Sasaki T, Ihara H, Idota Y, Yamagata H, Udaka S (1985) Efficient synthesis and secretion of a thermophilic α-amylase by protein-producing *Bacillus brevis* 47 carrying the *Bacillus stearothermophilus* amylase gene. J Bacteriol 164:1182-1187

Udaka S (1976) Extracellular production of proteins by microorganism. Part I. screening for protein producing bacteria. Agric Biol Chem 40:523-528

Yamada H, Tsukagoshi N, Udaka S (1981) Morphological alterations of cell wall concomitant with protein release in a protein-producing bacterium, *Bacillus brevis* 47. J Bacteriol 148:322-332

Yamagata H, Adachi T, Tsuboi A, Takao M, Sasaki T, Tsukagoshi N, Udaka S (1987) Cloning and characterization of the 5′ region of the cell wall protein gene operon in *Bacillus brevis* 47. J Bacteriol 169:1239-1245

Permebility Properties and the Use of S-Layers for the Production of Ultrafiltration Membranes

M. Sára, G. Wolf and U.B. Sleytr

Zentrum für Ultrastrukturforschung
Universität für Bodenkultur
Gregor Mendel Str. 33
A-1180 Wien
AUSTRIA

Since S-layers represent the outermost cell wall component in many prokaryotes and completely cover the cell surface during cell growth and cell division it was suggested that, among other functions, they could act as protective coats against hostile macromolecules always present in natural environments (Sleytr 1978, Sleytr and Messner 1983). For various mesophilic bacteria it was suggested that S-layers can protect the cell wall proper from lysozyme attack (Beveridge 1981, Nermut and Murray 1967, Sleytr and Messner 1983, Wallinder and Neujahr 1971), although in most cases the resistance of the murein sacculus itself towards lysozyme has never been studied. Among a large number of the highly thermophilic *Bacillus stearothermophilus* strains possessing S-layers of different structure not a single strain was found to be insensitive to lysozyme (Messner et al. 1984). Derived from its molecular size (3.0 nm x 3.0 nm x 4.5 nm) it is evident that the pores passing through these S-layer lattices must exceed 3.0 nm to allow free passage.

In contrast to high resolution electron microscope studies and computer image reconstruction procedures providing information on the mass distribution in crystalline lattices, permeability studies using well-defined molecules will lead to information on the "functional" pores size of S-layers under well-defined environmental conditions. Since muramidases with progressively increasing molecular weights are not available, the exclusion limits of S-layer lattices were determined by the "space technique" which measures passive solute uptake through porous structures (Scherrer and Gerhardt 1971). For this purpose the proper permeability barrier, the plasma membrane, was extracted by mild detergent treatment under conditions that preserved the integrity of the peptidoglycan-containing layer and the S-layer (Sára and Sleytr 1987a). Nucleic acid-containing material was removed by nuclease treatment, followed by a number of wash cycles. The preparations obtained by this procedure are referred to as "native sacculi" in the following. In this case the permeability barrier consisted of three layers, namely the outer S-layer, the peptidoglycan-containing layer, and an incomplete inner S-layer, attached to the rigid cell wall layer.

After crosslinking the native sacculi with glutaraldehyde, the peptidoglycan could be removed by lysozyme treatment without destroying the structural integrity of the native sacculi. Such preparations are referred to as "glutaraldehyde-treated" sacculi. The permeability barrier consisted of the outer S-layer and the incomplete inner S-layer. The distance between them corresponded approximately to the thickness of the originally present peptidoglycan-containing layer.

Native and glutaraldehyde-treated sacculi were then used for the permeability studies. For this purpose 1 g of the wet pellet was mixed with 1 ml of solutions of either mannose, myoglobin (17,000), carbonic anhydrase (30,000), ovalbumin (43,000), bovine serum albumin (67,000) or dextran T-510 (450,000) and incubated for 1 h at 20°C. After sedimentation of the sacculi, the concentration of the solutions in the respective supernatant was determined. The intercellular aqueous phase of the pellet was determined with dextran T-510 which is too large to pass through pores in the S-layer lattices. The "space values" for a given solute were calculated as follows (Scherrer and Gerhardt, 1971):

$$\% \; S_{sol} = V_s/W_p \; (C_o/C_f - 1) \; . \; 100$$

V_s represents the volume of the solution added to the pellet, W_p is the weight of the pellet. C_o is the initial solute concentration, and C_f is the final solute concentration as measured in the supernatant after equilibration and repacking the pellet. By substraction of the contents of the intercellular phase (S_{in}; determined with dextran T-510) the uptake of a given solute into the interior of the sacculi can be calculated. This is expressed by the corrected uptake value R_{sol} for a given solute.

$$\% \; R_{sol} = (S_{sol} - S_{in}) \; (100 - S_{in})$$

The maximum permeable space was determined with mannose. As shown in Fig. 1, myoglobin and carbonic anhydrase could pass through the pores in the S-layer lattices of *Bacillus stearothermophilus* strains NRS 1536/3c and PV72. Ovalbumin could partly penetrate, whereas bovine serum albumin was rejected. If carbonic anhydrase with a molecular size of 4.1 nm x 4.1 nm x 4.7 nm (Squire and Himmel 1979) penetrates in lengthwise orientation, it is evident that the "functional" pore size in both S-layer lattices must exceed 4.1 nm. Despite the structural difference of the S-layers of the two *Bacillus stearothermophilus* strains (Sleytr et al., 1986) a steep increase in solute rejection occurred between 30,000 and 45,000 molecular weight. These findings confirmed that the molecular sieving properties of S-layers are not determined by the lattice structure and composition of S-layers. It seems more likely that bacteria growing under comparable selection pressure are endowed with pores similar in size and function.

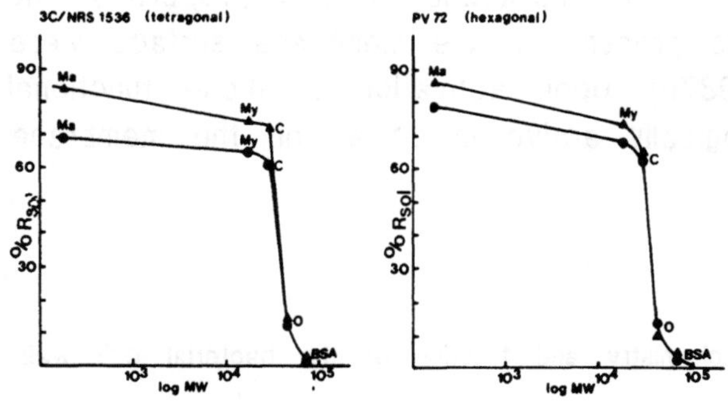

Fig.1. Percentage uptake of solutes (%R_{sol}) as a function of molecular weight by native sacculi (●—●) and glutaraldehyde-treated sacculi (▲—▲) of *Bacillus stearothermophilus* (a) strain NRS 1536/3c and (b) PV72. Ma, mannose; My, myoglobin; C, carbonic anhydrase; O, ovalbumin; BSA, bovine serum albumin

From high-resolution electron microscope and permeability studies it is evident that S-layers work in the range of synthetic ultrafiltration membranes, which cover a pore size from approximately 1-20 nm (Strathmann 1981). When isolated S-layers were deposited on microporous supports by a pressure dependent procedure and subsequently treated with bifunctional crosslinkers, stable "composite membranes" were obtained (Sleytr and Sára 1986, Sára and Sleytr 1987b). The active ultrafiltration layer consisted of coherent S-layer monolayers. For the characterization of such S-layer ultrafiltration membranes (SUM) parameters were used, as developed for ultrafiltration membranes produced of synthetic polymers. The *water flux* of SUM was in the range of 150 - 200 l/m^2/h, when measured with particle free, distilled water at 0.2 M Pa.

The performance of ultrafiltration membranes is usually derived from their *rejection curves.* For this purpose the rejection coefficients for test molecules with progressively increasing molecular weights are determined and plotted versus the log of their molecular weights. When myoglobin, carbonic anhydrase, ovalbumin and bovine serum albumin were used as testmolecules, a steep increase in solute rejection was observed to occur between 30,000 and 45,000 molecular weight using SUM made of S-layer material of the highly thermophilic Bacillaceae. The percentile rejection coefficients for carbonic anhydrase were 0-5 %. Ovalbumin was retained from 80 to 90 %, and for bovine serum albumin the rejection coefficients were >96%.

Concerning the chemical stability, SUM were resistent towards a broad spectrum of organic solvents, and between pH 1 - 13. Upon crosslinking with glutaraldehyde, SUM exhibited a net negative charge originating from carboxyl groups present on the S-layer surface (Sára and Sleytr 1987b). For altering the membrane surface properties, the carboxyl- and hydroxyl groups present on the membrane surface were modified (Sára and Sleytr 1987b). Upon activation of these functional groups, immobilization of biologically active molecules on the membrane surface is also possible.

References

Beveridge TJ (1981) Ultrastructure, chemistry and function of the bacterial cell wall. Int Rev Cytol 72:229-317

Messner P, Hollaus F, Sleytr UB (1984) Paracrystalline cell wall surface layers of different *Bacillus stearothermophilus* strains. Int J Syst Bacteriol 34:202-210

Nermut MV, Murray RGE (1967) Ultrastructure of the cell wall of *Bacillus polymyxa.* J Bacteriol 93:1949-1965

Sára M, Sleytr UB (1987a) Molecular-sieving through S-layers of *Bacillus stearothermophilus* strains. J Bacteriol 169:4092-4098

Sára M, Sleytr UB (1987b) Production and characteristics of ultrafiltration membranes with uniform pores from two-dimensional arrays of proteins. J Membr Sci 33:27-49

Scherrer R, Gerhardt P (1971) Molecular-sieving by the *Bacillus megaterium* cell wall and protoplast. J Bacteriol 107:718-735

Sleytr UB (1978) Regular arrays of macromolecules on bacterial cell walls: structure, chemistry, assembly, and function. Int Rev Cytol 53:1-64

Sleytr UB, Messner P (1983) Crystalline surface layers on bacteria. Annu Rev Microbiol 37:311-339

Sleytr UB, Sára (1986) Ultrafiltration membranes with uniform pores from crystalline bacterial cell envelope layers. Appl Microbiol Biotechnol 25:83-90

Sleytr UB, Sára M, Küpcü Z, Messner P (1986) Structural and chemical characterization of the S-layers of selected strains of *Bacillus stearothermophilus* and *Desulfotomaculum nigrificans.* Arch Microbiol 149:19-24

Squire PG, Himmel ME (1979) Hydrodynamics and protein hydration. Arch Biochem Biophys 196:165-177

Strathmann H (1981) Membrane separation process. J Membr Sci 10:121-189

Wallinder IB, Neujahr HY (1971) Cell wall and peptidoglycan from *Lactobacillus fermenti.* J Bacteriol 105:918-926

Comparative Studies on Synthetic and S-Layer Ultrafiltration Membranes

C. Manigley, G. Wolf, M. Sára and U.B. Sleytr

Zentrum für Ultrastrukturforschung
Universität für Bodenkultur
Gregor Mendel Str. 33
A-1180 Wien
AUSTRIA

The paper describes comparative studies on S-layer ultrafiltration membranes (SUM) and commercial ultrafiltration membranes. In S-layer ultrafiltration membranes (SUM) the permselective layer consists of coherent S-layer fragments, in which protein- or glycoprotein molecules are arranged according to a crystalline lattice (Sára and Sleytr 1987). By contrast, the permselective skin in commercial ultrafiltration membranes showing a thickness of 0.1 um or less and an amorphous structure, is prepared of polymers as regenerated cellulose, cellulose acetate or polysulfone (Strathmann 1981).

According to the definition suggested by the European Society of Membrane Science and Technology ultrafiltration is a membrane operation in which the membrane fractionates components of a liquid, predominantly according to molecular size and shape. Ultrafiltration is used for the separation of molecules with molecular weights ranging from 500 to 300,000.

For describing the performance of ultrafiltration membranes the following parameters are used:

- flux
- rejection coefficients
- rejection curve
- nominal molecular weight cut off *(NMWCO)*
- adsorption
- chemical stability

For the comparative studies the following ultrafiltration membranes were chosen:

Manufacturer	Type	Membrane Material	NMWCO
Amicon	YM 30	regenerated cellulose	30,000
Amicon	PM 30	polysulfone	30,000
Millipore	PTGC	polysulfone	10,000
Millipore	PTTK	polysulfone	30,000
Sartorius	SM 145	cellulose acetate	10,000
Sartorius	SM 145	cellulose acetate	20,000
DDS	GR 60PP	polysulfone	25,000
DDS	FS 50PP	fluorinated polymer	30,000
DDS	CA 600PP	cellulose acetate	20,000
DDS	GS 61PP	polysulfone	20,000
Nuclepore	Type F		20,000
Nuclepore	Type F		50,000
Nuclepore	Type C		20,000
Nuclepore	Type C		50,000
Thomapor			50,000

Two ultrafiltration membranes made of polyhydantoin showing a nominal molecular weight cut off of 25,000 and 50,000, respectively, and which have been developed by Bayer AG (Leverkusen, FRG), are not yet commercially available.

Flux

The flux is the volume passing through the membrane per unit area and unit time at a defined pressure. It is usually given in $l/m^2/h$. For flux determination the use of distilled, particle free water is recommended.

Rejection Coefficients

Conventionally, the performance of ultrafiltration membranes is derived from the rejection coefficients (%R) obtained from examination of a series of compounds of progressively increasing molecular weights. For this purpose the ultrafiltration cell was charged with 4 ml of a test solution (0.8 mg/ml distilled water) of a well characterized solute. After volume reduction by factor 4 at a pressure of 0.2 M Pa the concentration of the solute in the retentate and in the permeate (3 ml) was determined. Cytochrome c (11,000), myoglobin (17,000), carbonic anhydrase (30,000), ovalbumin (43,000) and bovine serum albumin (67,000) were used as testmolecules. The rejection coefficients were calculated as described previously (Blatt 1976).

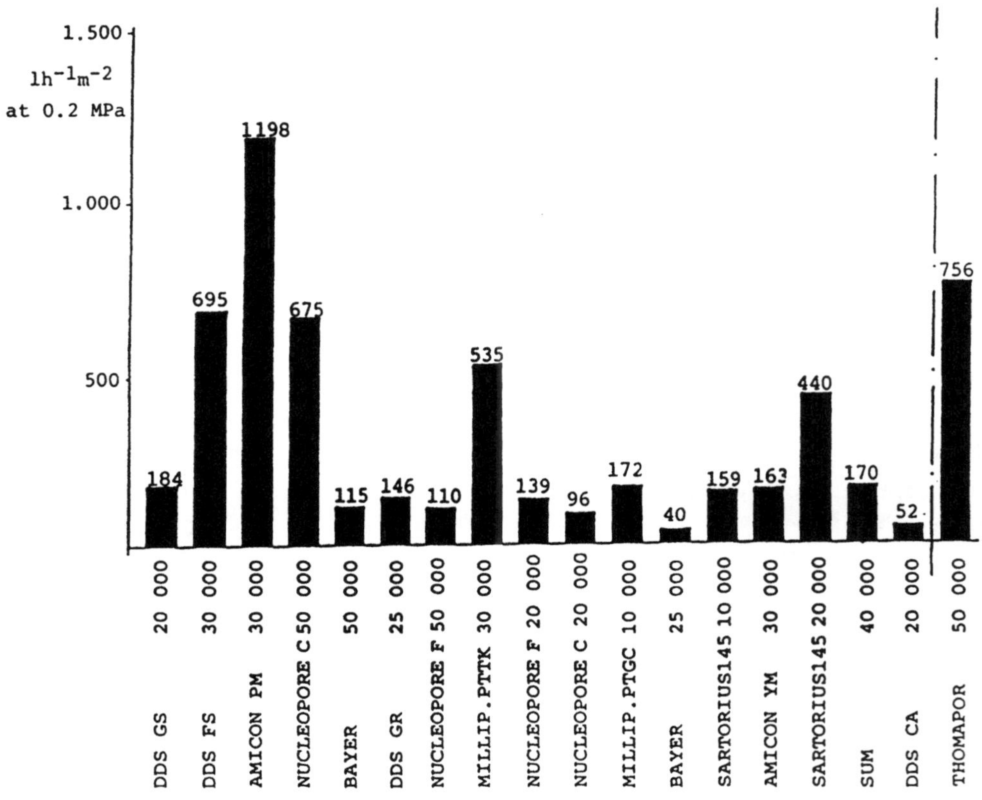

Fig.1. Flux of various ultrafiltration membranes, measured with distilled, particle free water.

Rejection curve

The percentile rejection coefficients measured for the various testmolecules were plotted versus the log of their molecular weights (Fig.3). Theoretically, an ideal membrane with a narrow distribution of pore diameters would display a sharp rejection curve.

The performance of the ultrafiltration membranes was also derived from their ability to fractionate macrosolute mixtures into classes of graded molecular size. The macrosolute mixture was composed of myoglobin (17,000), carbonic anhydrase (30,000), ovalbumin (43,000) and bovine serum albumin (67,000). The feed solution (2.2 mg protein/ml) was concentrated by factor 4. Feed solution, permeate and retentate were analyzed by SDS-PAGE (Fig.3).

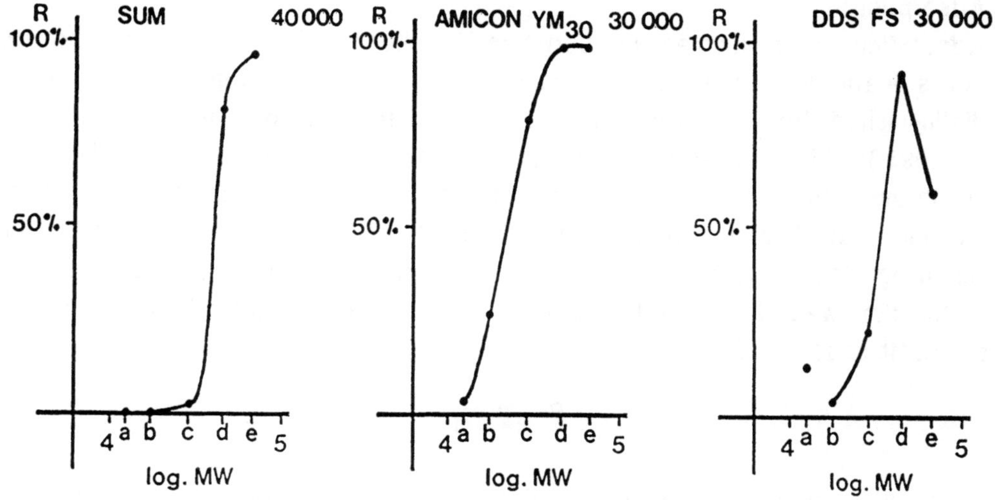

Fig.2. Rejection curves of various ultrafiltration membranes. The percentile rejection coefficients (%R) are plotted versus the log of the molecular weights of the testmolecules. (a) cytochrome c (11,000), (b) myoglobin (17,000), (c) carbonic anhydrase (30,000), (d) ovalbumin (43,000), (e) bovine serum albumin (67,000)

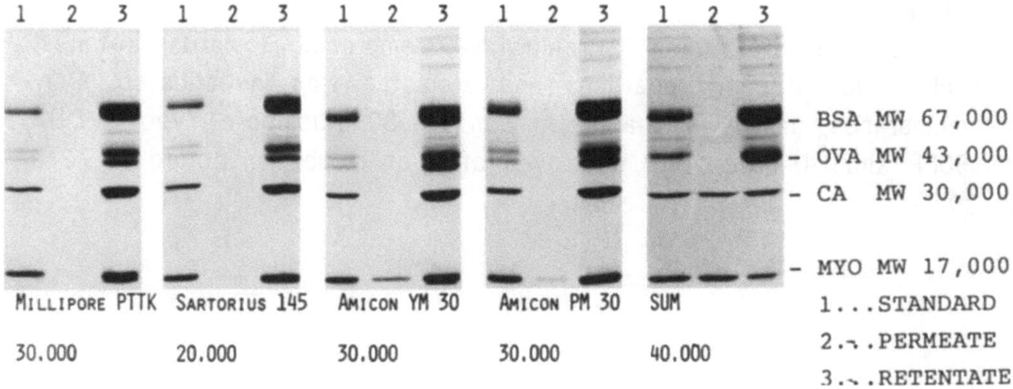

Fig.3. Separation of a protein mixture using different ultrafiltration membranes.

Nominal Molecular Weight Cut Off

For evaluating ultrafiltration membranes, an arbitrary characterization in terms of the nominal molecular weight cut off (NMWCO) is universally used as a convenient parameter. The NMWCO corresponds to the molecular weight showing 95% rejection.

Adsorption

Adsorption of molecules or particles to the membrane surface and its pores leads to a decrease in membrane performance. Generally it can be distinguished between reversible and irreversible adsorption (Reihanian et al. 1983). For determination of irreversibly adsorbed protein, the membrane surface was exposed to bovine serum albumin (5 mg/ml distilled water) and incubated for 2 h at 20°C without pressure. After removing the protein solution and washing the membrane surface the water flux was determined. The relative flux was calculated as described by Matthiasson (1983):

$$RF = \frac{J_0}{J_A}$$

J_0 is the initial water flux; J_A is the water flux after exposure of the membrane surface in the bovine serum albumin solution. 1-RF expresses the flux losses.

The results indicate that not only the membrane material, but also other factors like the porosity, the pore size distribution and the surface roughness can influence membrane adsorption.

Chemical Resistance

The resistance of various ultrafiltration membranes towards ketones, alcohols, chlorinated hydrocarbons and aromates was investigated. Only six membranes, namely Thomapor, Amicon YM 30, Nuclepore Type F, DDS FS 50PP and SUM were not destroyed after an incubation period of 72 h at 20°C.

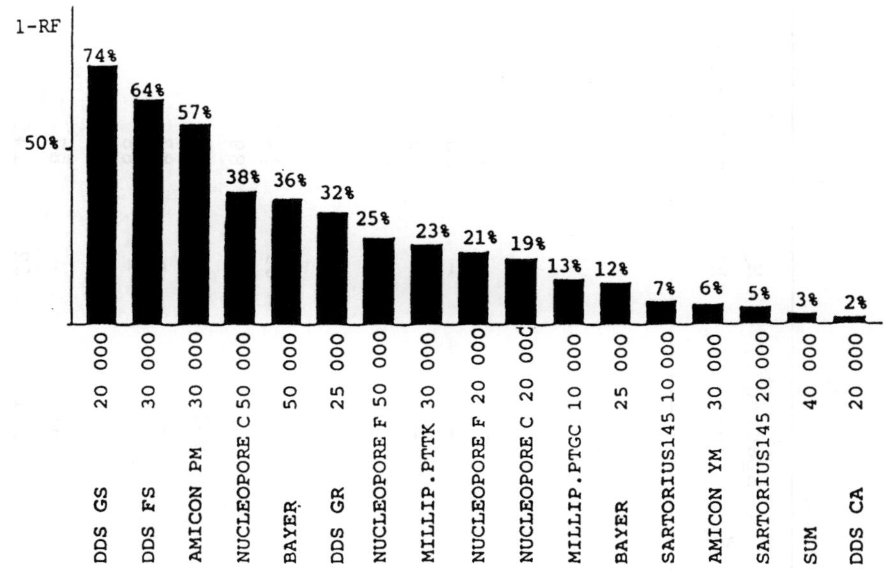

Fig.4. Relative flux losses (1-RF) of various ultrafiltration membranes after exposure to the bovine serum albumin solution

References

Blatt WF (1976) Principles and practice of ultrafiltration. In: Meares P (ed) Membrane separation processes. Elsevier, Amsterdam, pp 81-120

Matthiasson E (1983) The role of macromolecule adsorption in fouling of ultrafiltration membranes. J Membr Sci 16:23-36

Reihanian H, Robertson CR, Michaels AS (1983) Mechanisms of polarization and fouling of ultrafiltration membranes by proteins. J Membr Sci 16:237-258

Sára M, Sleytr UB (1987) Production and characteristics of ultrafiltration membranes with uniform pores from two-dimensional arrays of proteins. J Membr Sci 33:27-49

Strathmann H (1981) Membrane separation processes. J Membr Sci 10:121-189

Crystalline surface layers on bacteria

Uwe B. Sleytr and Paul Messner

Organism	Lattice[a]	Spacing[b] (nm)	M_r (kD)	References
Gram-negative eubacteria[c]				
Spirochetes				
Spirochaeta plicatilis	P	-	-	23
Spirochaeta stenostrepta Z1	P	-	-	80
Spirochaeta zuelzera	H	-	-	89
Spirochaeta litoralis R1	P	-	-	78
Spirochaeta aurantia J1	P	-	-	25
Treponema pallidum Nichols	P	-	-	96
Treponema phagedenis, biotype Reiter	P	-	69	147
Treponema refringens	H	-	-	89
Treponema minutum (CIP 5162)	H	8-10	-	87
Treponema calligyrum (CIP 6441)	H	-	-	87
Treponema genitalis VDRL-2	H	-	-	88
Treponema microdentium	P	-	-	136
Treponema microdentium (4 strains)	H	~10	-	87
Treponema sp. E-21	H	16.3	62	145

Aerobic/microaerophilic, motile, helical/vibroid bacteria				
Aquaspirillum serpens VHA	H	14.5	140	27,28,45[h],64,117,118
Aquaspirillum serpens MW5[g]	H	16	150,125	112,219
Aquaspirillum serpens MW6	H	22.5	-	160
Aquaspirillum putridiconchylium (ATCC 15279)	O	11.8/7.8	156	13,220
Aquaspirillum metamorphum (ATCC 15280)[g]	S	5/10	-	14
	H	20	-	
Aquaspirillum "Ordal"[g]	H	20	-	15,82
	S	12	-	
Aquaspirillum sp.	H	12-14	-	83
Campylobacter fetus 23D	H	-	98	150,252
Campylobacter pylori	H	-	-	M.R.Dickson, pers commun.
Aerobic rods and cocci				
Pseudomonadaceae				
Pseudomonas putida	S	-	-	49
Pseudomonas acidovorans (46 strains)	S	6.5	-	126,127
Pseudomonas acidovorans (7 strains)	H	8	-	126,127
Pseudomonas acidovorans (3 strains)	S	12	-	126,127
Pseudomonas acidovorans (DSM 39)	S	11	130	35[h]
Pseudomonas delafieldii	S	6.5	-	126,127
Pseudomonas facilis	S	6.5	-	126,127
Pseudomonas avenae (NCPPB 1011)	S	6.7	-	245

Continued on following page

continued

Organism	Lattice[a]	Spacing[b] (nm)	M_r (kD)	References
Azotobacteriaceae				
Azotobacter vinelandii	S	~12.5	65	19,20,21[h],22,186
Azomonas agilis	H	10	-	85
Azomonas insigne	H	12.5	-	65
Methylococcaceae				
Methylomonas albus	H	-	-	75,97
Neisseriaceae				
Acinetobacter sp. MJT/F5/199A	S	6/8	65	193,230
Acinetobacter sp. MJT/F5/5	H	11	-	66,200
Other genera				
Thermus aquaticus	S	-	-	77
Thermomicrobium roseum (ATCC 27502)	H	-	75	151,168
Lampropedia hyalina[g]	H H	23 13.5	240 31.5	5,37,164
Facultatively anaerobic rods				
Vibrionaceae				
Aeromonas salmonicida A450	S	5	49	105,167,234
Aeromonas salmonicida V75/93	S	10	54	51,52,53
Aeromonas hydrophila	S	12	52	46
Other genera				
Cardiobacterium hominis	S	5.5	-	176

Anaerobic straight, curved and helical rods				
Bacteroidaceae				
Bacteroides buccae (ATCC 33574)	P	-	-	70
Bacteroides buccae	H	21.5	-	191[h]
Bacteroides nodosus (6 strains)	H	6-7	-	56
Bacteroides capillus (ATCC 33690, ATCC 33691)	P	-	-	70
Bacteroides pentasaceus NP333 and WPH61	P	-	-	70
Bacteroides forsythus (ATCC 43037)	P	-	-	226
Bacteroides heparinolyticus	H	20	-	106
Bacteroides sp. ES42 and ES57	P	-	-	226
Bacteroides sp.	H	-	-	169
Wolinella recta	H	20	-	124
Selenomonas palpitans	H	13.5	-	237
Rickettsias and Chlamydias				
Rickettsiaceae				
Rickettsia prowazekii	S	13	-	163
Rickettsia akari	S	13	-	162
Chlamydiaceae				
Chlamydia trachomatis TE55	H	17.5	40	36[h]
Chlamydia psittaci	H	18	14-44	140
Phototrophic bacteria[d]				
Rhodospirillaceae				
Rhodospirillum rubrum	H	16.5	-	122,182

Continued on following page

continued

Organism	Lattice[a]	Spacing[b] (nm)	M_r (kD)	References
Rhodospirillum molischianum	S	10	-	63
Rhodospirillum salexigens (DSM 2132)	H	-	68	55
Rhodopseudomonas palustris	H	16	-	63
Rhodopseudomonas acidophila	H	-	-	227
Chromatiaceae				
Chromatium okenii	G,H	19	-	72
Chromatium weissei	G,H	19	-	72
Chromatium warmingii	G,H	~30	-	71
Chromatium buderi	G,H	~35	-	172,174
Chromatium gracile	H	5	-	174
Thiocapsa floridana 9314	S	8	-	224
Amoebobacter bacillosus	H	4	-	40
Ectothiorhodospira mobilis 8112	S	6-8	-	175
Ectothiorhodospira halochloris BN9850	P	-	-	94
Chlorobiaceae				
Pelodictyon sp.	P	-	-	32
Chlorochromatium aggregatum	P	-	-	32
Gliding bacteria[d]				
Myxococcaceae				
Myxococcus xanthus DK1050	only biochem. evidence		74	139

Cytophagaceae				
Cytophaga johnsonae (ATCC 17061)	G,H	-	-	30,165
Flexibacter columnaris	G,H	-	-	30,165
Flexibacter polymorphus (ATCC 27820)	G,H	25	13-80	30,165,178,179
Budding and/or appandaged bacteria[d]				
prosthecate halophilic microorganism (hyphomicrobium type)	T	9.0	-	109
Pedomicrobium sp.	P	18.5	-	62
Caulobacter crescentus CB15	H	23.5	130	205,206,207
Caulobacter sp.	P	15	-	44
Planctomyces sp. ES	H	12	-	187
"*Planctomyces gracilis*" Hortobágyi 1965[#]	H	-	-	212
Chemolithotrophic bacteria[d]				
Nitrosomonas sp.	H	15	-	241
Thiobacillus kabobis	P	-	-	177
Genus not yet classified:				
Acetogenium kivui (ATCC 33488)	H	20	-	131
Acetogenium kivui (DSM 2030)	H	19	~90	170
Nitrosocystis oceanus[g]	H	12	-	173,242
	S	7.5	-	
Pelobacter carbinolicus (DSM 2909)	H	22	-	48
Thermotoga maritima (DSM 3109)	H	12.4	-	93

Continued on following page

continued

Organism	Lattice[a]	Spacing[b] (nm)	M_r (kD)	References
Gram-positive eubacteria[e]				
Gram-positive cocci				
Deinococcus radiodurans Sark	H	16.5	115	125,229
Deinococcus radiodurans R1	H	18	104	7,8,9[h],121,201, 255
Peptostreptococcus asaccharolyticus (ATCC 14963)	H	14.2	91	W.Buckel & P.Messner, unpubl.
Peptostreptococcus magnus AHC 5155	P	-	-	137
Endospore-forming rods and cocci				
Bacillus subtilis	S	7.5	-	91
Bacillus alvei 183	H	10-12	-	257
Bacillus anthracis	H	7-10	-	47,61,81
Bacillus brevis 47[g]	H H	14.5 14.5	150 130	231,232,233,256
Bacillus brevis 47-5(phage-resistant mutant)	H	14.5	150	231,233
Bacillus brevis S1	H	17	129	2
Bacillus cereus (ATCC 4342)	S	9-10	-	50
Bacillus coagulans Hammer, E38-66	O	9.4/7.3	100	M.Sára, unpubl.
Bacillus fastidiosus	S	12-13.5	-	81,129
Bacillus megaterium	S	19	-	183
Bacillus polymyxa (NCIB 4747)	S	10.0	-	31,59,67,161

Bacillus schlegelii (DSM 2000)	S	9	-	185
Bacillus sphaericus (NTCC 9602, wild type)	S	13	142	74
Bacillus sphaericus (NTCC 9602, lmw, spontaneous variant)	S	13	120	74
Bacillus sphaericus P-1	S	13	140	3,90,132,133[h]
Bacillus sphaericus (several strains)	S	13	-	81,192,194
Bacillus sphaericus (several strains)	S	-	127-133	135,254
Bacillus stearothermophilus (39 strains)	S,O,H	5.2-17.9	83-170	154,197
Bacillus stearothermophilus (4 strains)	S,O,H	8.5-22.5	93-170	202
Bacillus aneurinolyticus				
type I (7 strains)	S	9.6	129	1
type II (13 strains)	S	9.6	114	1
Bacillus macroides A and D[g]	S S	5-6 9-10	- -	81
Bacillus psychrophilus W16A	S	15-16	-	81
Bacillus sp. (CIP 76-111)	H	11	255	130
Bacillus sp. KL1	P	-	-	238
Bacillus sp. M3198	P	-	-	69
Bacillus sp.	P	-	107,103	73
Clostridium aceticum (DSM 1496)	S	10	-	24
Clostridium aceticum (DSM 1496)	S	~12	-	253[h]
Clostridium botulinum	P	12.5	-	225
Clostridium difficile GAI 0714	S	8.1	45,32	148
Clostridium difficile (9 strains)	S	-	46-32	104
Clostridium formicoaceticum (DSM 912)	S	10	-	G.Acker, unpubl.
Clostridium novyi	P	10	-	184,243

Continued on following page

continued

Organism	Lattice[a]	Spacing[b] (nm)	M_r (kD)	References
Clostridium polysaccharolyticum	P	-	-	236
Clostridium sporogenes	S	-	-	11
Clostridium symbiosum HB25	H	~17	-	116, P.Messner, unpubl.
Clostridium tetani	P	14.5	-	225
Clostridium thermoautotrophicum	S	-	-	246
Clostridium thermohydrosulfuricum (13 strains)	H	~14	-	79,194,198
Clostridium thermohydrosulfuricum L111-69	H	14	140	41,196,234
Clostridium thermohydrosulfuricum (DSM 568)	H	-	-	34[h]
Clostridium thermosaccharolyticum (3 strains)	S	8-9.5	-	79,198
Clostridium thermosaccharolyticum D120-70	S	11	140	41,195,196
Clostridium thermosaccharolyticum (DSM 571)	S	-	-	33[h]
Clostridium thermolacticum (DSM 2911)	P	7	-	134
Clostridium tyrobutyricum (ATCC 25755 and 3 strains)	P	-	110-125	10,76
Clostridium lentoputrescens (ATCC 17791)	O	11	140	U.B.Sleytr, unpubl
Clostridium tartarivorum (2 strains)	S	10	-	79,194
Desulfotomaculum nigrificans (several strains)	S	8-12	-	86,199
Desulfotomaculum nigrificans (4 strains)	S,O	9.0-14.6	85-131	202
Sporosarcina ureae (ATCC 13881)	S	12.8	150	12,51[h],81,218
Regular, nonsporing, Gram-positive rods				
Lactobacillus acidophilus (ATCC 4357)	P	-	43	101

Lactobacillus acidophilus (ATCC 4356)	P	-	-,46	171
Lactobacillus helveticus (ATCC 10797)	P	-	51	101
Lactobacillus helveticus, biotype jogurti (ATCC 521)	P	-	-	146
Lactobacillus casei	H	-	-	6
Lactobacillus brevis (ATCC 8287)	O	7.0/4.5	51	142,143
Lactobacillus brevis (YIT 0017)	P	-	-	146
Lactobacillus buchneri (ATCC 4005)	H	6	55	144
Lactobacillus buchneri (YIT 0040)	P	-	-	146
Lactobacillus fermentum (NCTC 7230)	O	9.6/6.2	52	101,103
Lactobacillus bulgaricus (YIT 0045)	P	-	51.5	146
Lactobacillus sp.	S	10.7	-	240
Irregular, nonsporing, Gram-positive rods				
Corynebacterium diphteriae C4[g]	S	5.3	-	102
	S	3.5	-	
Eubacterium tenue	H	15.7	-	240
Eubacterium yurii, subsp. yurii	P	-	-	141
Eubacterium lentum AHP 6099	P	-	-	137
Eubacterium sp. AHN 990	H	15.6	-	137
Acetobacterium woodii WB1	O	~12/8	-	149
Thermoanaerobacter ethanolicus	H	-	-	247

Continued on following page

continued

Organism	Lattice[a]	Spacing[b] (nm)	M_r (kD)	References
Cyanobacteria				
Microcystacae				
Aphanothece halophytica (ATCC 29534)	P	-	200	190
Cyanothece minerva	S	-	-	68
Synechococcus sp. GC	P	-	-	235
Synechocystis aquatilis	H	10	-	204
Synechocystis aquatilis f. *salina*	H	-	-	203
Synechocystis fuscopigmentosa	H	-	-	203
Synechocystis sp. CB3	H	15.5	-	138
Synechocystis sp. CLII	H	15.2	100	100[h]
Synechocystis sp. PCC 6803	P	-	-	235
Microcystis firma	P	-	-	235
Microcystis inearta	H	11.3	-	203
Microcystis marginata	H	~14	-	107
Xenococcacae				
Chroococcidiopsis sp.	P	-	-	29
Chroococcacean cyanobacteria (22 strains)	H	-	-	235

Archaebacteria[f]				
Halobacteriales				
Halobacterium halobium, several strains	H	13-16	200	84,111,180,223, 248
Halobacterium salinarium 1 (DSM 668, ATCC 19700)	H	-	200	152,153,158
Halobacterium cutirubrum	H	15-16.5	200	42,123; H.König & K.O.Stetter,unpubl
Halobacterium volcanii DS2 (ATCC 29605)	H	13-14	190	38,39,159
Halobacterium volcanii (CCM 3361)	P	-	185	H.König & K.O. Stetter, unpubl.
Halobacterium volcanii	H	14	200	110
Halobacterium saccharovorum (DSM 1137)	P	-	220	H.König & K.O. Stetter, unpubl.
Halobacterium sp. r-4 (DSM 1411)	P	-	-	H.König & K.O. Stetter, unpubl.
Halobacterium sp. 5	H	-	-	213
Methanobacteriales				
Methanobacterium sp. G2R	P	-	-	211
Methanothermus fervidus V24S (DSM 2088)	H	10.8	93	111,217; H.König & P.Messner, unpubl.
Methanothermus sociabilis KF 1/F1	H	19.2	-	128, H.König & P.Messner, unpubl.
Methanococcales				
Methanococcus vannielii SB (DSM 1224)	H	10.8	60	98,115
Methanococcus voltae PS (DSM 1537)	H	~10	76	115,119
Methanococcus thermolithotrophicus SN1 (DSM 2095)	P	-	83	92,115

Continued on following page

continued

Organism	Lattice[a]	Spacing[b] (nm)	M_r (kD)	References
Methanococcus jannaschii JAL-1 (DSM 2661)	H	-	90,82	99,115
Methanococcus aeolicus PL-15/H	H	12.3	-	P.Messner & K.O. Stetter, unpubl.
Methanomicrobiales				
Methanomicrobium mobile BP	P	-	-	166; H.König, pers. commun.
Methanogenium cariaci JR1 (DSM 1497)	H	14	117	181,258
Methanogenium marisnigri JR1 (DSM 1498)	H	-	138	181,258
Methanogenium tatii (DSM 2702)	H	-	120	258
Methanogenium thermophilicum (DSM 2640)	H	-	130	259
Methanospirillum hungatei JF1 (DSM 864, ATCC 27890)	B	2.8	-	16,221
Methanospirillum hungatei GP1 (DSM 1101)	B	2.8	-	189[h]
Methanoplanus limicola M3 (DSM 2279)	H	14.0	143	249
Methanosarcina mazei (DSM 2053)	P	-	-	4
Methanosarcina acetivorans C2A (DSM 2834, ATCC 35395)	P	-	-	209
Methanothrix soehngenii (DSM 2139)	B	2.8	-	115,260
Methanothrix concilii	B	2.8	-	17,18
Methanolobus tindarius Tindari 3 (DSM 2278)	H	11.2	156	114,115
Methanolobus siciliae T 4/M	P	-	-	K.O.Stetter & P.Messner, unpubl.
Methanococcoides methylutens TMA-10 (ATCC 33938)	P	-	-	210

Methanocorpusculum parvum XII (DSM 3823)	H	14.3	90	261; G.Zellner & P.Messner, unpubl.
Sulfolobales				
Sulfolobus brierleyi (DSM 1651)	H	20.0	123	115,263; M.Sára, unpubl.
Sulfolobus acidocaldarius (ATCC 27360)	H	20	-	244
Sulfolobus acidocaldarius 98-3 (DSM 639)	H	22	140-170	43[h],156,228
Sulfolobus acidocaldarius 7	H	20	100,40	95
Sulfolobus solfataricus-"*Caldariella acidophila*" (DSM 1616, DSM 1617)	H	20	-	157,263
Acidianus infernus So4a (DSM 3191)	H	12.4	-	188, K.O.Stetter & P.Messner, unpubl.
Thermodiscus maritimus S2	P	-	-	60
Pyrodictium occultum PL-19 (DSM 2709)	H	16;20.5	172	214,216; P.Messner unpubl.
Pyrodictium brockii S1 (DSM 2708)	H	23.1	150	216; K.O.Stetter, unpubl.
Staphylothermus marinus F1	H	-	-	58
Pyrococcus furiosus Vc 1 (DSM 3638)	H	12.5	-	57
Thermoproteales				
Thermoproteus tenax Kra1 (DSM 2078)	H	31.4	-	115,155,250[h],264
Thermoproteus neutrophilus Hvv24 (DSM 2338)	H	30.3	-	60,155
Thermoproteus autotrophicus	H	27.7	-	W.Zillig, pers. commun.; M.Sára, unpubl.
Thermoproteus sp. H3	H	-	-	60

Continued on following page

continued

Organism	Lattice[a]	Spacing[b] (nm)	M_r (kD)	References
Thermofilum pendens Hvv3 (DSM 2475)	H	-	-	265
Desulfurococcus mobilis (DSM 2161)	S	-	-	251[h],266
Desulfurococcus mucosus (DSM 2162)	S	17.5	-	266; M.Sára, unpubl.
Thermococcus celer (DSM 2476)	P	-	-	262
Genus not yet classified:				
"*Archaeococcus*" -like bacterium, strain Z	H	17.5	132	G.Zellner & P.Messner, unpubl.
"Square bacterium" (halophilic)	H,S	14-23	-	108,222,239

[a]Type of lattice: H, hexagonal - T, trimers; S, square; O, oblique; B, bands of structural units; P, periodic structures not further characterized (e.g. evidence from thin sections).
[b]Center-to-center spacing.
[c]Classification according to ref. 120.
[d]Classification according to ref. 26.
[e]Classification according to ref. 208.
[f]Classification according to Stetter (215) und König (113).
[g]Two superimposed S-layers.
[h]Three-dimensional image reconstruction has been accomplished.

References

1. Abe M, Kimoto M (1984) Distribution of two types of regular-array particles in the cell wall of *Bacillus aneurinolyticus* (Kimura et Aoyama). Microbiol Immunol 28:841-846
2. Abe M, Kimoto M, Yoshii Z (1983) Structural and chemical characterization of macromolecular arrays in the cell wall of *Bacillus brevis* S1. FEMS Microbiol Lett 18:263-267
3. Aebi U, Smith PR, Dubochet J, Henry C, Kellenberger E (1973) A study of the structure of the T-layer of *Bacillus brevis.* J Supramolec Struct 1:498-522
4. Aldrich HC, Robinson RW, Williams DS (1986) Ultrastructure of *Methanosarcina mazei.* System Appl Microbiol 7:314-319
5. Austin JW, Murray RGE (1988) The surface layers of *Lampropedia hyalina.* This volume
6. Barker DC, Thorne KJI (1970) Spheroplasts of *Lactobacillus casei* and the cellular distribution of bactoprenol. J Cell Sci 7:755-785
7. Baumeister W, Kübler O (1978) Topographic study of the cell surface of *Micrococcus radiodurans.* Proc Natl Acad Sci USA 75:5525-5528
8. Baumeister W, Karrenberg F, Rachel R, Engel A, ten Heggeler B, Saxton WO (1982) The major cell envelope protein of *Micrococcus radiodurans* (R1). Structural and chemical characterization. Eur J Biochem 125:535-544
9. Baumeister W, Barth M, Hegerl R, Guckenberger R, Hahn M, Saxton WO (1986) The three-dimensional structure of the regular surface layer (HPI-layer) of *Deinococcus radiodurans.* J Mol Biol 187:241-253
10. Bergère J-L, Hayes H, Commissaire J (1986) Major protein components in the cell envelope of *Clostridium tyrobutyricum.* Ann Inst Pasteur/Microbiol 137 B:271-282
11. Betz JV (1970) Sheathed cells in cultures of *Clostridium sporogenes.* J Bacteriol 103:814-825
12. Beveridge TJ (1979) Surface arrays on the cell wall of *Sporosarcina ureae.* J Bacteriol 139:1039-1048
13. Beveridge TJ, Murray RGE (1974) Superficial macromolecular arrays on the cell wall of *Spirillum putridiconchylium.* J Bacteriol 119:1019-1038
14. Beveridge TJ, Murray RGE (1975) Surface arrays on the cell wall of *Spirillum metamorphum.* J Bacteriol 124:1529-1544
15. Beveridge TJ, Murray RGE (1976) Superficial cell wall layers on *Spirillum* "Ordal" and their in vitro reassembly. Can J Microbiol 22:567-582
16. Beveridge TJ, Stewart M, Doyle RJ, Sprott GD (1985) Unusual stability of the *Methanospirillum hungatei* sheath. J Bacteriol 162:728-737
17. Beveridge TJ, Harris BJ, Patel GB, Sprott GD (1986) Cell division and filament splitting in *Methanothrix concilii.* Can J Microbiol 32:779-786
18. Beveridge TJ, Patel GB, Harris BJ, Sprott GD (1986) The ultrastructure of *Methanothrix concilii,* a mesophilic aceticlastic methanogen. Can J Microbiol 32: 703-710
19. Bingle WH, Doran JL, Page WJ (1984) Regular surface layer of *Azotobacter vinelandii.* J Bacteriol 159:251-259
20. Bingle WH, Doran JL, Page WJ (1986) Characterization of the surface layer protein from *Azotobacter vinelandii.* Can J Microbiol 32:112-120
21. Bingle WH, Engelhardt H, Page WJ, Baumeister W (1987) Three-dimensional structure of the regular tetragonal surface layer of *Azotobacter vinelandii.* J Bacteriol 169:5008-5015
22. Bingle WH, Whippey PW, Doran JL, Murray RGE, Page WJ (1987) Structure of the *Azotobacter vinelandii* surface layer. J Bacteriol 169:802-810
23. Blakemore RP, Canale-Parola E (1973) Morphological and ecological characteristics of *Spirochaeta plicatilis.* Arch Mikrobiol 89:273-289
24. Braun M, Mayer F, Gottschalk G (1981) *Clostridium aceticum* (Wieringa), a microorganism producing acetic acid from molecular hydrogen and carbon dioxide. Arch Microbiol 128:288-293

25. Breznak JA, Canale-Parola E (1969) *Spirochaeta aurantia,* a pigmented, facultatively anaerobic spirochete. J Bacteriol 97:386-395
26. Buchanan RE, Gibbons NE (eds) (1974) Bergey's manual of determinative bacteriology, 8th ed. Williams and Wilkins, Baltimore
27. Buckmire FLA, Murray RGE (1970) Studies on the cell wall of *Spirillum serpens.* 1. Isolation and partial purification of the outermost cell wall layer. Can J Microbiol 16:1011-1022
28. Buckmire FLA, Murray RGE (1976) Substructure and in vitro assembly of the outer, structured layer of *Spirillum serpens.* J Bacteriol 125:290-299
29. Büdel B, Rhiel E (1985) A new cell wall structure in a symbiotic and a free-living strain of the blue-alga genus *Chroococcidiopsis (Pleuracapsales).* Arch Microbiol 143:117-121
30. Burchard RP (1981) Gliding motility of prokaryotes: Ultrastructure, physiology, and genetics. Annu Rev Microbiol 35:497-529
31. Burley SK, Murray RGE (I983) Structure of the regular surface layer of *Bacillus polymyxa.* Can J Microbiol 29:775-780
32. Caldwell DE, Tiedje JM (1975) A morphological study of anaerobic bacteria from the hypolimia of two Michigan lakes. Can J Microbiol 21:362-376
33. Cejka Z, Baumeister W (1987) Three-dimensional structure of the surface protein of *Clostridium thermosaccharolyticum.* FEMS Microbiol Lett 44:13-18
34. Cejka Z, Hegerl R, Baumeister W (1986) Three-dimensional structure of the surface layer protein of *Clostridium thermohydrosulfuricum.* J Ultrastruct Mol Struct Res 96:1-11
35. Chalcroft JP, Engelhardt H, Baumeister W (1986) Three-dimensional structure of a regular surface layer from *Pseudomonas acidovorans.* Arch Microbiol 144:196-200
36. Chang J-J, Leonard K, Arad T, Pitt T, Zhang Y-X, Zhang L-H (1982) Structural studies on the outer envelope of *Chlamydia trachomatis* by electron microscopy. J Mol Biol 161:579-590
37. Chapman JA, Murray RGE, Salton MRJ (1963) The surface anatomy of *Lampropedia hyalina.* Proc R Soc London Sect B 158:498-513
38. Cohen S, Oren A, Shilo M (1983) The divalent cation requirement of Dead Sea halobacteria. Arch Microbiol 136:184-190
39. Cohen S, Shilo M, Kessel M (1983) Ultrastructural study of the cell wall of *Halobacterium volcanii* from the Dead Sea. Ultramicroscopy 11:330-331
40. Cohen-Bazire G, Kunisawa R, Pfennig N (1969) Comparative study of the structure of gas vacuoles. J Bacteriol 100:1049-1061
41. Crowther RA, Sleytr UB (1977) An analysis of the fine structure of the surface layer from two strains of *Clostridia,* including correction for distorted images. J Ultrastruct Res 58:41-49
42. D'Aoust J-Y, Kushner DJ (1972) The regular hexagonal surface layer of *Halobacterium cutirubrum:* a honeycomb network. Can J Microbiol 18:1767-1768
43. Deatherage JF, Taylor KA, Amos LA (1983) Three-dimensional arrangement of the cell wall protein of *Sulfolobus acidocaldarius.* J Mol Biol 167:823-852
44. De Boer WE, Spit BJ (1964) A new type of bacterial cell wall structure revealed by replica technique. Antonie van Leeuwenhoek J Microbiol Serol 30:239-248
45. Dickson MR, Downing KH, Wu WH, Glaeser RM (1986) Three-dimensional structure of the surface layer protein of *Aquaspirillum serpens* VHA determined by electron crystallography. J Bacteriol 167:1025-1034
46. Dooley JSG, Murray RGE, Trust TJ (1988) The S-layers of pathogenic strains of *Aeromonas hydrophila.* This volume
47. Doyle RJ, Beveridge TJ, Stewart M, Ezzell E (1986) Evidence for a surface array on *Bacillus anthracis.* Abstr 86th Annu Meeting Am Soc Microbiol, Washington DC, 191 J-18
48. Dubourguier H-C, Samain E, Prensier G, Albagnac G (1986) Characterization of two strains of *Pelobacter carbinolicus* isolated from anaerobic digesters. Arch Microbiol 145:248-253
49. Dupoy G (1973) Performance and applications of the Toulouse 3 million volt electron microscope. J Microsc 97:3-28

50. Ellar DJ, Lundgren DG (1967) Ordered substructure in the cell wall of *Bacillus cereus.* J Bacteriol 94:1778-1780
51. Engelhardt H, Saxton WO, Baumeister W (1986) Three-dimensional structure of the tetragonal surface layer of *Sporosarcina ureae.* J Bacteriol 168:309-317
52. Evenberg D, Lugtenberg B (1982) Cell surface of the fish pathogenic bacterium *Aeromonas salmonicida.* II. Purification and characterization of a major cell envelope protein related to autoagglutination, adhesion and virulence. Biochim Biophys Acta 684:249-254
53. Evenberg D, Versluis R, Lugtenberg B (1985) Biochemical and immunological characterization of the cell surface of the fish pathogenic bacterium *Aeromonas salmonicida.* Biochim Biophys Acta 815:233-244
54. Evenberg D, van Boxtel R, Lugtenberg B, Schurer F, Blommaert J, Bootsma R (1982) Cell surface of the fish pathogenic bacterium *Aeromonas salmonicida.* I. Relationship between autoagglutination and the presence of a major cell envelope protein. Biochim Biophys Acta 684:241-248
55. Evers D, Weckesser J, Drews G (1984) Protein on the cell surface of the moderately halophilic phototrophic bacterium *Rhodospirillum salexigens.* J Bacteriol 160: 107-111
56. Every D, Skerman TM (1980) Ultrastructure of the *Bacteroides nodosus* cell envelope layers and surface. J Bacteriol 141:845-857
57. Fiala G, Stetter KO (1986) *Pyrococcus furiosus* sp. nov. represents a novel genus of marine heterotrophic archaebacteria growing optimally at 100°C. Arch Microbiol 145:56-61
58. Fiala G, Stetter KO, Jannasch HW, Langworthy TA, Madon J (1986) *Staphylothermus marinus* sp. nov. represents a novel genus of extremely thermophilic submarine heterotrophic archaebacteria growing up to 98°C. System Appl Microbiol 8:106-113
59. Finch JT, Klug A, Nermut MV (1967) The structure of the macromolecular units on the cell walls of *Bacillus polymyxa.* J Cell Sci 2:587-590
60. Fischer F, Zillig W, Stetter KO, Schreiber G (1983) Chemolithoautotrophic metabolism of anaerobic extremely thermophilic archaebacteria. Nature 301:511-513
61. Gerhardt P (1967) Cytology of *Bacillus anthracis.* Fed Proc 26:1504-1517
62. Ghiorse WC, Hirsch P (1979) An ultrastructural study of iron and manganese deposition associated with extracellular polymers of *Pedomicrobium* -like budding bacteria. Arch Microbiol 123:213-226
63. Giesbrecht P (1969) Die wichtigsten Feinstrukturen der Bakterien im elektronenmikroskopischen Bild. In: Henneberg G (ed) Bildatlas pathogener Mikroorganismen, Bd. III. Fischer, Stuttgart, pp 1-141
64. Glaeser RM, Chiu W, Grano D, Taylor K (1980) Morphological model of the surface-layer array in *Spirillum serpens.* In: Baumeister W, Vogell W (eds) Electron microscopy at molecular dimensions. Springer, Berlin, pp 22-26
65. Glauert AM (1962) Fine structure and function in bacteria. Proc 5th Int Congr Electron Microsc, Philadelphia, 2:RR-1
66. Glauert AM, Thornley MJ (1971) Fine structure and radiation resistance in *Acinetobacter:* a comparison of a range of strains. J Cell Sci 8:19-41
67. Goundry J, Davison AL, Archibald AR, Baddiley J (1967) The structure of the cell wall of *Bacillus polymyxa* (N.C.I.B. 4747). Biochem J 104:1c-2c
68. Gromov BV, Gavrilova OV, Konovalov ES (1986) Ultrastruktura kletok cianobakterij roda *Cyanothece.* Mikrobiol 55: 821-824
69. Haapasalo M, Lounatmaa K, Ranta H (1988) Paracrystalline surface of a novel *Bacillus* sp. M3198. This volume
70. Haapasalo M, Lounatmaa K, Ranta H, Shah H, Ranta K (1985) Ultrastructure of *Bacteroides capillus, B. buccae, B. pentasaceus, B. oris, B. oralis, B. veroralis* and pentose-sugar-fermenting *Bacteroides* sp. from humans with periapical osteitis: occurrence of external proteinaceous cell wall layer. Int J Syst Bacteriol 35:65-72
71. Hageage GJ, Gherna RL (1970) Electron microscopy of the cell envelope of *Chromatium warmingii.* Bact Proc 57

72. Hageage GJ Jr, Gherna RL (1971) Surface structure of *Chromatium okenii* and *Chromatium weissei*. J Bacteriol 106:687-690
73. Haikara A, Lantto R, Mäkinen V, Lounatmaa K (1985) Ropiness of beer caused by *Bacillus* sp. Proc 20th Congr Eur Brewery Convention, Helsinki, 443-450
74. Hastie AT, Brinton CC Jr (1979) Isolation, characterization and *in vitro* self-assembly of the tetragonally arranged layer of *Bacillus sphaericus*. J Bacteriol 138: 999-1009
75. Haubold R (1978) Two different types of surface structures of methane utilizing bacteria. Z Allg Mikrobiol 18:511-515
76. Hayes H, Commissaire J, Bergère J-L (1984) A high-molecular-mass cell wall protein released from *Clostridium tyrobutyricum* by heat treatment. FEBS Lett 174: 284-288
77. Heinen UJ, Heinen W (1972) Characteristics and properties of a caldo-active bacterium producing extracellular enzymes and two related strains. Arch Mikrobiol 82:1-23
78. Hespell RB, Canale-Parola E (1970) *Spirochaeta litoralis* sp. n., a strictly anaerobic marine spirochete. Arch Mikrobiol 74:1-18
79. Hollaus F, Sleytr UB (1972) On the taxonomy and fine structure of some hyperthermophilic saccharolytic clostridia. Arch Mikrobiol 86:129-146
80. Holt SC, Canale-Parola E (1968) Fine structure of *Spirochaeta stenostrepta*, a free-living, anaerobic spirochete. J Bacteriol 96:822-835
81. Holt SC, Leadbetter ER (1969) Comparative ultrastructure of selected aerobic spore forming bacteria: a freeze-etching study. Bacteriol Rev 33:346-378
82. Holt SC, Beveridge TJ (1982) Electron microscopy: its development and application to microbiology. Can J Microbiol 28:1-53
83. Houwink AL (1953) A macromolecular monolayer in the cell wall of *Spirillum* spec.. Biochim Biophys Acta 10:360-366
84. Houwink AL (1956) Flagella, gas vacuoles and cell-wall structure in *Halobacterium halobium*; an electron microscope study. J Gen Microbiol 15:146-150
85. Houwink AL (1963) In: Brieger EM (ed) Structure and ultrastructure of microorganisms. Academic Press, New York, p 282
86. Houwink AL (1963) In: Brieger EM (ed) Structure and ultrastructure of microorganisms. Academic Press, New York, p 294
87. Hovind-Hougen K (1974) The ultrastructure of cultivable treponemes. 2. *Treponema calligyrum, Treponema minutum* and *Treponema microdentium*. Acta Pathol Microbiol Scand Sect B 82:495-507
88. Hovind-Hougen K (1975) The ultrastructure of cultivable treponemes. 3. *Treponema genitalis*. Acta Pathol Microbiol Scand Sect B 83:91-99
89. Hovind-Hougen K (1976) Determination by means of electron microscopy of morphological criteria of value for classification of some *Spirochetes*, in particular *Treponemes*. Acta Pathol Microbiol Scand Sect B Suppl 255:1-41
90. Howard L, Tipper DJ (1973) A polypeptide bacteriophage receptor: modified cell wall protein subunits in bacteriophage-resistant mutants of *Bacillus sphaericus* strain P-1. J Bacteriol 113:1491-1504
91. Howatson AF, Russel WC (1964) Observations on cell-wall structure in *Bacillus subtilis*. J Appl Physics 35:3086-3087
92. Huber H, Thomm M, König H, Thies G, Stetter KO (1982) *Methanococcus thermolithotrophicus*, a novel thermophilic lithotrophic methanogen. Arch Microbiol 132: 47-50
93. Huber R, Langworthy TA, König H, Thomm M, Woese CR, Sleytr UB, Stetter KO (1986) *Thermotoga maritima* sp. nov. represents a new genus of unique extremely thermophilic eubacteria growing up to 90°C. Arch Microbiol 144:324-333
94. Imhoff JF, Trüper HG (1977) *Ectothiorhodospira halochloris* sp. nov., a new extremely halophilic phototrophic bacterium containing bacteriochlorophyll b. Arch Microbiol 114:115-121
95. Inatomi K, Ohba M, Oshima T (1983) Chemical properties of proteinaceous cell wall from acidothermophilic, *Sulfolobus acidocaldarius*. Chem Lett 1191-1194

96. Jackson S, Black SH (1971) Ultrastructure of *Treponema pallidum* Nichols following lysis by physical and chemical methods. 1. Envelope, wall, membrane and fibrils. Arch Mikrobiol 76:308-324
97. Jeffries P, Wilkinson JF (1978) Electron microscopy of the cell wall complex of *Methylomonas albus.* Arch Microbiol 119:227-229
98. Jones JB, Bowers B, Stadtman TC (1977) *Methanococcus vannielii:* ultrastructure and sensitivity to detergents and antibiotics. J Bacteriol 130:1357-1363
99. Jones WJ, Leigh JA, Mayer F, Woese CR, Wolfe RS (1983) *Methanococcus jannaschii* sp. nov., an extremely thermophilic methanogen from a submarine hydrothermal vent. Arch Microbiol 136:254-261
100. Karlsson B, Vaara T, Lounatmaa K, Gyllenberg H (1983) Three-dimensional structure of the regularly constructed surface layer from *Synechocystis* sp strain CLII. J Bacteriol 156:1338-1343
101. Kawata T (1981) Electron microscopic studies on surface structures of bacterial cells with special reference to regular arrays on bacterial surfaces. Electron Microsc 15:115-122
102. Kawata T, Masuda K (1972) Extracellular crystalline lattice material of *Corynebacterium diphteriae* revealed by electron microscopy. Jpn J Microbiol 16:515-523
103. Kawata T, Masuda K, Yoshino K, Fujimoto M (1974) Regular array in the cell wall of *Lactobacillus fermenti* as revealed by freeze-etching and negative staining. Jpn J Microbiol 18:469-476
104. Kawata T, Takeoka A, Takumi K, Masuda K (1984) Demonstration and preliminary characterization of a regular array in the cell wall of *Clostridium difficile.* FEMS Microbiol Lett 24:323-328
105. Kay WW, Phipps BM, Ishiguro EE, Olafson RW, Trust TJ (1984) Surface layer virulence A-protein from *Aeromonas salmonicida* strains. Can J Biochem Cell Biol 62:1064-1071
106. Kerosuo E, Haapasalo M, Lounatmaa K, Ranta H (1988) Ultrastructural comparison of *Bacteroides heparinolyticus* and *Bacteroides buccae.* This volume
107. Kessel M (1978) A unique crystalline wall layer in the cyanobacterium *Microcystis marginata.* J Ultrastruct Res 62:203-212
108. Kessel M, Cohen Y (1982) Ultrastructure of square bacteria from a brine pool in Southern Sinai. J Bacteriol 150:851-860
109. Kessel M, Radermacher M, Frank J (1985) The structure of the surface layer of a brine pond micro-organism: correlation averaging applied to a double layered lattice structure. J Microsc 139:63-74
110. Kessel M, Cohen S, Buhle EL Jr, Aebi U (1986) Structure of the cell wall of a moderately halophilic bacterium from the Dead Sea. Proc 11th Int Congr Electron Microsc Kyoto 3:2471-2472
111. Kirk RG, Ginzburg M (1972) Ultrastructure of two species of *Halobacterium.* J Ultrastruct Res 41:80-94
112. Kist ML, Murray RGE (1984) Components of the regular surface array of *Aquaspirillum serpens* MW5 and their assembly in vitro. J Bacteriol 157:599-606
113. König H (1986) Biologie der Archaebakterien. Biologie in unserer Zeit 16:71-82
114. König H, Stetter KO (1982) Isolation and characterization of *Methanolobus tindarius,* sp. nov., a coccoid methanogen growing only on methanol and methylamines. Zentralbl Bakteriol Hyg I Abt Orig C3:478-490
115. König H, Stetter KO (1986) Studies on archaebacterial S-layers. System Appl Microbiol 7:300-309
116. König H, Buckel W, Langworthy TA (1985) Ultrastructure of the cell envelope and amino acid composition of the murein of *Clostridium symbiosum.* FEMS Microbiol Lett 30:283-288
117. Koval SF, Murray RGE (1981) Cell wall proteins of *Aquaspirillum serpens.* J Bacteriol 146:1083-1090
118. Koval SF, Murray RGE (1985) Effect of calcium on the *in vivo* assembly of the surface protein of *Aquaspirillum serpens* VHA. Can J Microbiol 31:261-267
119. Koval SF, Jarrell KF (1987) Ultrastructure and biochemistry of the cell wall of *Methanococcus voltae.* J Bacteriol 169:1298-1306

120. Krieg NR, Holt JG (eds) (1984) Bergey's manual of systematic bacteriology, vol 1. Williams and Wilkins, Baltimore
121. Kübler O, Engel A, Zingsheim HP, Emde B, Hahn M, Heisse W, Baumeister W (1980) Structure of the HPI-layer of *Micrococcus radiodurans*. In: Baumeister W, Vogell W (eds) Electron microscopy at molecular dimensions. Springer, Berlin, pp 11-21
122. Kuhn PJ, Holt SC (1972) Characterisation of a blue mutant of *Rhodospirillum rubrum*. Biochim Biophys Acta 261:267-275
123. Kushner DJ, Bayley ST, Boring J, Kates M, Gibbons NE (1964) Morphological and chemical properties of cell envelopes of the extreme halophile, *Halobacterium cutirubrum*. Can J Microbiol 10:483-497
124. Lai CH, Listgarten MA, Tanner ACR, Socransky SS (1981) Ultrastructures of *Bacteroides gracilis, Campylobacter concius, Wolinella recta,* and *Eikenella corrodens,* all from humans with periodontal disease. Int J Syst Bacteriol 31:465-475
125. Lancy P Jr, Murray RGE (1977) The envelope of *Micrococcus radiodurans:* isolation, purification and preliminary analysis of the wall layers. Can J Microbiol 24:162-176
126. Lapchine L (1976) Ultrastructure de la paroi de *Pseudomonas acidovorans*. J Microscopie Biol Cell 25:67-72
127. Lapchine L (1979) Regularly arranged structures on the surface of some *Pseudomonas* sp. FEMS Microbiol Lett 5:223-225
128. Lauerer G, Kristjansson JK, Langworthy TA, König H, Stetter KO (1986) *Methanothermus sociabilis* sp. nov., a second species within the *Methanothermaceae* growing at 97°C. System Appl Microbiol 8:100-105
129. Leadbetter ER, Holt SC (1968) The fine structure of *Bacillus fastidiosus*. J Gen Microbiol 52:299-307
130. Leduc M, Rousseau M, van Heijenoort J (1977) Structure of the cell wall of *Bacillus* species C.I.P. 76-111. Eur J Biochem 80:153-163
131. Leigh JA, Mayer F, Wolfe RS (1981) *Acetogenium kivui,* a new thermophilic hydrogen-oxidizing, acetogenic bacterium. Arch Microbiol 129:275-280
132. Lepault J, Pitt T (1984) Projected structure of unstained, frozen-hydrated S-layer of *Bacillus brevis*. EMBO J 3:101-105
133. Lepault J, Martin N, Leonard K (1986) Three-dimensional structure of the T-layer of *Bacillus sphaericus* P-1. J Bacteriol 168:303-308
134. LeRuyet P, Dubourguier H-C, Albagnac G, Prensier G (1985) Characterization of *Clostridium thermolacticum* sp. nov., a hydrolytic thermophilic anaerobe producing high amounts of lactate. System Appl Microbiol 6:196-202
135. Lewis LO, Yousten AA, Murray RGE (1987) Characterisation of the surface protein layers of the mosquito-pathogenic strains of *Bacillus sphaericus*. J Bacteriol 169:72-79
136. Listgarten MA, Socransky SS (1964) Electron microscopy of axial fibrils, outer envelope, and cell division of certain oral spirochetes. J Bacteriol 88:1087-1103
137. Lounatmaa K, Jousimies-Somer H, Grénman R, Rintala A (1988) Crystalline surface layers in anaerobic bacteria isolated from a patient with secretory otitis media and a draining ear. This volume
138. Lounatmaa K, Vaara T, Österlund K, Vaara M (1980) Ultrastructure of the cell wall of a *Synechocystis* strain. Can J Microbiol 26:204-208
139. Maeba PY (1986) Isolation of a surface glycoprotein from *Myxococcus xanthus*. J Bacteriol 166:644-650
140. Manire GP, Wyrick PB (1979) Cell envelopes of Chlamydiae: adaption for intracellular parasitism. In: Schlesinger D (ed) Microbiology-1979. American Society for Microbiology, Washington DC, pp 111-115
141. Margaret BS, Krywolap GN (1986) *Eubacterium yurii* subsp. *yurii* sp. nov. and *Eubacterium yurii* subsp. *Margaretiae* subsp. nov.: test tube brush bacteria from subgingival dental plaque. Int J Syst Bacteriol 36:145-149
142. Masuda K, Kawata T (1978) Ultrastructure and partial characterization of a regular array in the cell wall of *Lactobacillus brevis*. Microbiol Immunol 23:941-953

143. Masuda K, Kawata T (1980) Reassembly of the regularly arranged subunits in the cell wall of *Lactobacillus brevis* and their reattachment to cell walls. Microbiol Immunol 24:299-308
144. Masuda K, Kawata T (1981) Characterization of a regular array in the wall of *Lactobacillus buchneri* and its reattachment to the other wall components. J Gen Microbiol 124:81-90
145. Masuda K, Kawata T (1982) Isolation, properties, and reassembly of outer sheath carrying a polygonal array from an oral treponeme. J Bacteriol 150:1405-1413
146. Masuda K, Kawata T (1983) Distribution and chemical characterization of regular arrays in the cell walls of strains of the genus *Lactobacillus.* FEMS Microbiol Lett 20:145-150
147. Masuda K, Kawata T (1986) Isolation, and structural and chemical characterization of outer sheath carrying a polygonal array from *Treponema phagedenis* biotype Reiter. Microbiol Immunol 30:401-411
148. Masuda K, Kawata T (1986) Structural organization of regular arrays in the cell walls of *Clostridium difficile* strains. Proc 11th Int Congr Electron Microsc, Kyoto, 4:3365-3366
149. Mayer F, Lurz R, Schoberth S (1977) Electron microscopic investigation of the hydrogen-oxidizing acetate-forming anaerobic bacterium *Acetobacter woodii.* Arch Microbiol 115:207-213
150. McCoy EC, Doyle D, Burda K, Corbeil LB, Winter AJ (1975) Superficial antigens of *Campylobacter (Vibrio) fetus:* characterization of the antiphagocytic component. Infect Immun 11:517-525
151. Merkel GJ, Durham DR, Perry JJ (1980) The atypical cell wall composition of *Thermomicrobium roseum.* Can J Microbiol 26:556-559
152. Mescher MF, Strominger JL (1976) Structural (shape-maintaining) role of the cell surface glycoprotein of *Halobacterium salinarium.* Proc Natl Acad Sci USA 73: 2687-2691
153. Mescher MF, Strominger JL, Watson SW (1974) Protein and carbohydrate composition of the cell envelope of *Halobacterium salinarium.* J Bact 120:945-954
154. Messner P, Hollaus F, Sleytr UB (1984) Paracrystalline cell wall surface layers of different *Bacillus stearothermophilus* strains. Int J Syst Bacteriol 34:202-210
155. Messner P, Pum D, Sára M, Stetter KO, Sleytr UB (1986) Ultrastructure of the cell envelope of the archaebacteria *Thermoproteus tenax* and *Thermoproteus neutrophilus.* J Bacteriol 166:1046-1054
156. Michel H, Neugebauer D-Ch, Oesterhelt D (1980) The 2-d crystalline cell wall of *Sulfolobus acidocaldarius:* structure, solubilization, and reassembly. In: Baumeister W, Vogell W (eds) Electron microscopy at molecular dimensions. Springer, Berlin, pp 27-35
157. Millonig G, De Rosa M, Gambacorta A, Bu'lock JD (1975) Ultrastructure of an extremely thermophilic acidophilic microorganism. J Gen Microbiol 86:165-173
158. Mohr V, Larsen H (1963) On the structural transformations and lysis of *Halobacterium salinarium* in hypotonic and isotonic solutions. J Gen Microbiol 31:267-280
159. Mullakhanbhai MF, Larsen H (1975) *Halobacterium volcanii* spec. nov., a Dead Sea halobacterium with a moderate salt requirement. Arch Microbiol 104:207-214
160. Murray RGE (1968) Bacterial cell wall anatomy in relation to the formation of spheroplasts and protoplasts. In: Guze LB (ed), Microbial protoplasts, spheroplasts and L-forms. Williams and Wilkins, Baltimore, pp 1-18
161. Nermut MV, Murray RGE (1967) Ultrastructure of the cell wall of *Bacillus polymyxa.* J Bacteriol 93:1949-1965
162. Palmer EL, Martin ML, Mallavia L (1974) Ultrastructure of the surface of *Rickettsia prowazeki* and *Rickettsia akari.* Appl Microbiol 28:713-716
163. Palmer EL, Mallavia LP, Tzianabos T, Obijeski JF (1974) Electron microscopy of the cell wall of *Rickettsia prowazeki.* J Bacteriol 118:1158-1166
164. Pangborn J, Starr MP (1966) Ultrastructure of *Lampropedia hyalina.* J Bacteriol 91:2025-2030
165. Pate JL, Chang L-YE (1979) Evidence that gliding motility in prokaryotic cells is driven by rotary assemblies in the cell envelopes. Curr Microbiol 2:59-64

166. Paynter MB, Hungate RE (1968) Characterization of *Methanobacterium mobilis* sp. n., isolated from the bovine rumen. J Bacteriol 95:1943-1951
167. Phipps BM, Trust TJ, Ishiguro EE, Kay WW (1983) Purification and characterization of the cell surface virulent A protein from *Aeromonas salmonicida.* Biochemistry 22:2934-2939
168. Ramaley RF, Turner FR, Malick LE, Wilson RB (1978) The morphology and surface structure of some extremely thermophilic bacteria found in slightly alkaline hot springs. In: Friedman SM (ed) Biochemistry of thermophily. Academic Press, New York, pp 89-102
169. Ranta H, Lounatmaa K, Haapasalo M, Ranta K (1983) Isolation and ultrastructure of *Bacteroides* sp. with external cell-wall layer (S-layer) in periapical osteitis. Scand J Dent Res 91:458-464
170. Rasch M, Saxton WO, Baumeister W (1984) The regular surface layer of *Acetogenium kivui:* some structural, developmental and evolutionary aspects. FEMS Microbiol Lett 24:285-290
171. Ray B, Johnson MC (1986) Freeze-drying injury of surface layer protein and its protection in *Lactobacillus acidophilus.* Cryo-Letters 7:210-217
172. Remsen CC, Trüper HG (1973) The fine structure of *Chromatium buderi.* Arch Mikrobiol 90:269-280
173. Remsen CC, Valois FW, Watson SW (1967) Fine structure of the cytomembranes of *Nitrosocystis oceanus.* J Bacteriol 94:422-433
174. Remsen CC, Watson SW, Trüper HG (1970) Macromolecular subunits in the wall of marine photosynthetic bacteria. J Bacteriol 103:255-258
175. Remsen CC, Watson SW, Waterbury JB, Trüper HG (1968) Fine structure of *Ectothiorhodospira mobilis* Pelsh. J Bacteriol 95:2374-2392
176. Reyn A, Birch-Anderson A, Murray RGE (1971) The fine structure of *Cardiobacterium hominis.* Acta Pathol Microbiol Scand Sect B 79:51-60
177. Reynolds DM, Laishley EJ, Costerton JW (1981) Physiological and ultrastructural characterization of a new acidophilic *Thiobacillus* species *(T. kabobis).* Can J Microbiol 27:151-161
178. Ridgway HF (1977) Ultrastructural characterization of goblet-shaped particles from the cell wall of *Flexibacter polymorphus.* Can J Microbiol 23:1201-1213
179. Ridgway HF, Lewin RA (1973) Goblet-shaped subunits from the wall of a marine gliding microbe. J Gen Microbiol 79:119-128
180. Robertson JD, Schreil W, Reedy M (1982) *Halobacterium halobium.* 1. A thin-sectioning electron-microscopic study. J Ultrastruct Res 80:148-162
181. Romesser JA, Wolfe RS, Mayer F, Spiess E, Walther-Mauruschat A (1979) *Methanogenium,* a new genus of marine methanogenic bacteria, and characterization of *Methanogenium cariaci* sp. nov. and *Methanogenium marisnigri* sp. nov. Arch Microbiol 121:147-153
182. Salton MR, Williams RC (1954) Electron microscopy of the cell walls of *Bacillus megaterium* and *Rhodospirillum rubrum.* Biochim Biophys Acta 14:455-458
183. Sawatake M (1966) Macromolecular crystal-like arrays found in the cell wall of disrupted *Bacillus megatherium.* Proc 6th Int Congr Electron Microsc, Kyoto, 2: 279-280
184. Schallehn G, Wecke J (1974) Zur Feinstruktur der Zellwand von *Clostridium perfringens, Cl. septicum* und *Cl. novyi.* Zentralbl Bakteriol Hyg I Abt Orig A228:63-71
185. Schenk A, Aragno M (1979) *Bacillus schlegelii,* a new species of thermophilic, facultatively chemolithoautotrophic bacterium oxidizing molecular hydrogen. J Gen Microbiol 115:333-341
186. Schenk SF, Earhart CF (1981) Characterization of the predominant *Azotobacter vinelandii* envelope protein. J Bacteriol 146:398-403
187. Schmidt JM (1978) Isolation and ultrastructure of freshwater strains of *Planctomyces.* Curr Microbiol 1:65-70.

188. Segerer A, Neuner A, Kristjansson JK, Stetter KO (1986) *Acidianus infernus* gen. nov., sp. nov., and *Acidianus brierleyi* comb. nov.: facultatively aerobic, extremely acidophilic thermophilic sulfurmetabolizing archaebacteria. Int J Syst Bacteriol 36: 559-564

189. Shaw PJ, Hills GJ, Henwood JA, Harris JE, Archer DB (1985) Three-dimensional architecture of the cell sheath and septa of *Methanospirillum hungatei.* J Bacteriol 161:750-757

190. Simon RD (1981) Gliding motility in *Aphanothece halophytica:* analysis of wall proteins in *mot* mutants. J Bacteriol 148:315-321

191. Sjögren A, Hovmöller S, Farrants G, Ranta H, Haapsalo M, Ranta K, Lounatmaa K (1985) Structures of two different surface layers found in six *Bacteroides* strains. J Bacteriol 164:1278-1282

192. Sleytr U (1970) Gefrierätzung verschiedener Stämme von *Bacillus sphaericus.* Arch Mikrobiol 72:238-251

193. Sleytr UB, Thornely MJ (1973) Freeze-etching of the cell envelope of an *Acinetobacter* species which carries a regular array of surface subunits. J Bacteriol 116:1383-1397

194. Sleytr UB, Glauert AM (1975) Analysis of regular arrays of subunits on bacterial surfaces; evidence for a dynamic process of assembly. J Ultrastruct Res 50:103-116

195. Sleytr UB, Glauert (1976) Ultrastructure of the cell walls of two closely related clostridia that possess different regular arrays of surface subunits. J Bacteriol 126:869-882

196. Sleytr UB, Thorne KJI (1976) Chemical characterization of the regularly arrayed surface layers of *Clostridium thermosaccharolyticum* and *Clostridium thermohydrosulfuricum.* J Bacteriol 126:377-383

197. Sleytr U, Adam H, Klaushofer H (1967) Die elektronenmikroskopische Feinstruktur von Zellwand, Cytoplasmamembran und Geisseln von *Bacillus stearothermophilus,* dargestellt mit Hilfe der Gefrierätztechnik. Mikroskopie (Wien) 22:233-242

198. Sleytr U, Adam H, Klaushofer H (1968) Die Feinstruktur der Zellwandoberfläche von zwei thermophilen Clostridienarten, dargestellt mit Hilfe der Gefrierätztechnik. Mikroskopie (Wien) 23:1-10

199. Sleytr U, Adam H, Klaushofer H (1969) Die Feinstruktur der Zellwand und Cytoplasmamembran von *Clostridium nigrificans,* dargestellt mit Hilfe der Gefrierätz -und Ultradünnschnittechnik. Arch Mikrobiol 66:40-58

200. Sleytr UB, Thornley MJ, Glauert AM (1974) Location of the fracture faces within the cell envelope of *Acinetobacter* species strain MJT/F5/5. J Bacteriol 118:693-707

201. Sleytr UB, Kocur M, Glauert AM, Thornley MJ (1973) A study by freeze-etching of the fine structure of *Micrococcus radiodurans.* Arch Mikrobiol 94:77-87

202. Sleytr UB, Sára M, Küpcü Z, Messner P (1986) Structural and chemical characterization of S-layers of selected strains of *Bacillus stearothermophilus* and *Desulfotomaculum nigrificans.* Arch Microbiol 146:19-24

203. Šmarda J (1988) S-layer in cyanobacteria. This volume

204. Šmarda J, Čáslavská J, Komárek J (1979) Cell wall structure of *Synechocystis aquatilis* (Cyanophyceae). Arch Hydrobiol Suppl 56, Algological Studies 23:154-165

205. Smit J, Agabian N (1982) Cell surface patterning and morphogenesis: biogenesis of a periodic surface array during *Caulobacter* development. J Cell Biol 95:41-49

206. Smit J, Agabian N (1984) Cloning of the major protein of the *Caulobacter crescentus* periodic surface layer: detection and characterization of the cloned peptide by protein expression assays. J Bacteriol 160:1137-1145

207. Smit J, Grano DA, Glaeser RM, Agabian N (1981) A periodic surface array in *Caulobacter crescentus:* fine structure and chemical analysis. J Bacteriol 146:1135-1150

208. Sneath PHA, Mair NS, Sharpe ME, Holt JG (eds) (1986) Bergey's manual of systematic bacteriology, vol 2. Williams and Wilkins, Baltimore

209. Sowers KR, Baron SF, Ferry JG (1984) *Methanosarcina acetivorans* sp. nov., an acetotrophic methane-producing bacterium isolated from marine sediments. Appl Environ Microbiol 47:971-978
210. Sowers KR, Ferry JG (1983) Isolation and characterization of a methylotrophic marine methanogen, *Methanococcoides methylutens* gen. nov., sp. nov. Appl Environ Microbiol 45:684-690
211. Sprott GD, Sowden LC, Colvin JR, Jarrell KF, Beveridge TJ (1984) Methanogenesis in the absence of intracytoplasmic membranes. Can J Microbiol 30:594-604
212. Starr MP, Short KA, Schmidt JM (1984) Exclusion of the filamentous and rosette-forming bacterium *'Planctomyces gracilis'* Hortobágyi 1965 from the *Blastocaulis-Planctomyces* group. Int J Syst Bacteriol 34:465-469
213. Steensland H, Larsen H (1969) A study of the cell envelope of the halobacteria. J Gen Microbiol 55:325-336
214. Stetter KO (1982) Ultrathin mycelia-forming organisms from submarine volcanic areas having an optimum growth temperature of 105°C. Nature 300:258-260
215. Stetter KO (1985) Extrem thermophile Bakterien. Naturwissenschaften 72:291-301
216. Stetter KO, König H, Stackebrandt E (1983) *Pyrodictium* gen. nov., a new genus of submarine disc-shaped sulphur reducing archaebacteria growing optimally at 105°C. System Appl Microbiol 4:535-551
217. Stetter KO, Thomm M, Winter J, Wildgruber G, Huber H, Zillig W, Janékovic D, König H, Palm P, Wunderl S (1981) *Methanothermus fervidus*, sp. nov., a novel extremely thermophilic methanogen isolated from an icelandic hot spring. Zentralbl Bakteriol Hyg I Abt Orig C2:166-178
218. Stewart M, Beveridge TJ (1980) Structure of the regular surface layer of *Sporosarcina ureae*. J Bacteriol 142:302-309
219. Stewart M, Murray RGE (1982) Structure of the regular surface layer of *Aquaspirillum serpens* MW5. J Bacteriol 150:348-357
220. Stewart M, Beveridge TJ, Murray RGE (1980) Structure of the regular surface layer of *Spirillum putridiconchylium*. J Mol Biol 137:1-8
221. Stewart M, Beveridge TJ, Sprott GD (1985) Crystalline order to high resolution in the sheath of *Methanospirillum hungatei*: a cross-beta structure. J Mol Biol 183: 509-515
222. Stoeckenius W (1981) Walsby's square bacterium: fine structure of an orthogonal prokaryote. J Bacteriol 148:352-360
223. Stoeckenius W, Rowen R (1967) A morphological study of *Halobacterium halobium* and its lysis in media of low salt concentration. J Cell Biol 34:365-393
224. Takacs BJ, Holt SC (1971) *Thiocapsa floridana*; a cytological, physical and chemical characterization. I. Cytology of whole cells and isolated chromatophore membranes. Biochim Biophys Acta 233:258-277
225. Takagi A, Nakamura K, Ueda M (1965) Electron microscope studies of the intracytoplasmic membrane system in *Clostridium tetani* and *Clostridium botulinum*. Jpn J Microbiol 9:131-143
226. Tanner ACR, Listgarten MA, Ebersole JL, Strzempko MN (1986) *Bacteroides forsythus* sp. nov., a slow-growing, fusiform *Bacteroides* sp. from the human oral cavity. Int J Syst Bacteriol 36:213-221
227. Tauschel H-D, Hoehniger J (1974) The fine structure of *Rhodopseudomonas acidophila*. Can J Microbiol 20:13-17
228. Taylor KA, Deatherage JF, Amos LA (1982) Structure of the S-layer of *Sulfolobus acidocaldarius*. Nature 299:840-842
229. Thompson BG, Murray RGE, Boyce JF (1982) The association of the surface array and the outer membrane of *Deinococcus radiodurans*. Can J Microbiol 28:1081-1088
230. Thornley MJ (1975) Cell envelopes with regularly arranged surface subunits in *Acinetobacter* and related bacteria. CRC Crit Rev Microbiol 4:65-100
231. Tsuboi A, Tsukagoshi N, Udaka S. (1982) Reassembly *in vitro* of hexagonal surface arrays in a protein-producing bacterium, *Bacillus brevis* 47. J Bacteriol 151: 1485-1497

232. Tsuboi A, Uchihi R, Tabata R, Takahashi Y, Hasiba H, Sasaki T, Yamagata H, Tsukagoshi N, Udaka S (1986) Characterization of the genes coding for two major cell wall proteins from protein-producing *Bacillus brevis* 47: complete nucleotide sequence of the outer wall protein gene. J Bacteriol 168:365-373
233. Tsukagoshi N, Tabata R, Takemura T, Yamagata H, Udaka S (1984) Molecular cloning of a major cell wall protein gene from protein-producing *Bacillus brevis* 47 and its expression in *Escherichia coli* and *Bacillus subtilis*. J Bacteriol 158:1054-1060
234. Udey LR, Fryer JL (1978) Immunization of fish with bacterins of *Aeromonas salmonicida*. Marine Fisheries Rev 40:12-17
235. Vaara T (1982) The outermost surface structures in chroococcacean cyanobacteria. Can J Microbiol 28:929-941
236. Van Gylswyk NO, Morris EJ, Els HJ (1980) Sporulation and cell wall structure of *Clostridium polysaccharolyticum* comb. nov. (formerly *Fusobacterium polysaccharolyticum*). J Gen Microbiol 121:491-493
237. Van Iterson W (1956) Some thoughts on the possible relationship of bacterial flagella to cilia and flagella in other organisms. Proc 3rd Int Conf Electron Microsc, London, 602-607
238. Wahlberg J, Tynkkynen S, Taylor N, Uotila J, Kuusinen K, Kuusinen M, Häggblom M, Viljanen J, Villstedt R, Nurmiaho-Lassila E-L, Lounatmaa K (1987) A novel type of cell wall structure with two periodic layers (S-layer) of a *Bacillus* sp. strain KL1. FEMS Microbiol Lett 40:75-80
239. Walsby AE (1980) A square bacterium. Nature 283:69-71
240. Wang DN, Sjögren A, Hovmöller S, Lounatmaa K (1986) Structure studies of S-layers by electron microscopy. Proc 11th Int Congr Electron Microsc, Kyoto, 3: 2435-2436
241. Watson SW, Remsen CC (1969) Macromolecular subunits in the walls of marine nitrifying bacteria. Science 163:685-686
242. Watson SW, Remsen (1970) Cell envelope of *Nitrosocystis oceanus*. J Ultrastruct Res 33:148-160
243. Wecke J, Reinicke B, Schallehn G (1974) Remarkable differences in the ultrastructure of the cell wall of toxigenic clostridia. Proc 8th Int Congr Electron Microsc, Canberra, 2:644-645
244. Weiss RL (1974) Subunit cell wall of *Sulfolobus acidocaldarius*. J Bacteriol 118: 275-284
245. Wells B, Horne RW, Lund BM, King NR (1983) The ultrastructure of *Pseudomonas avenae* I. Paracrystalline surface layer and extracellular material. Micron 14:11-28
246. Wiegel J, Braun M, Gottschalk G (1981) *Clostridium thermoautotrophicum* species novum, a thermophile producing acetate from molecular hydrogen and carbon dioxide. Curr Microbiol 5:255-260
247. Wiegel J, Ljungdahl LG (1981) *Thermoanaerobacter ethanolicus* gen. nov., spec. nov., a new, extreme thermophilic, anaerobic bacterium. Arch Microbiol 128:343-348
248. Wieland F, Lechner J, Sumper M (1982) The cell wall glycoprotein of halobacteria: structural, functional and biosynthetic aspects. Zentralbl Bakteriol Hyg I Abt Orig C3:161-170
249. Wildgruber G, Thomm M, König H, Ober K, Ricchiuto T, Stetter KO (1982) *Methanoplanus limicola*, a plate-shaped methanogen representing a novel family, the Methanoplanaceae. Arch Microbiol 132:31-36
250. Wildhaber I, Baumeister W (1987) The cell envelope of *Thermoproteus tenax*: three-dimensional structure of the surface layer and its role in shape maintainance. EMBO J 6:1475-1480
251. Wildhaber I, Santarius U, Baumeister W (1987) Three-dimensional structure of the surface protein of *Desulfurococcus mobilis*. J Bacteriol 169:5563-5568
252. Winter AJ, McCoy EC, Fullmer CS, Burda K, Bier PJ (1978) Microcapsule of *Campylobacter fetus*: chemical and physical characterization. Infect Immun 22:963-971

253. Woodcock CLF, Engelhardt H, Baumeister W (1986) The tetragonal surface layer of *Clostridium aceticum:* three-dimensional structure and comparison with the hexagonal layer of *Clostridium thermohydrosulfuricum.* Eur J Cell Biol 42:211-217
254. Word NS, Yousten AA, Howard L (1983) Regularly structured and non-regularly structured surface layers of *Bacillus sphaericus.* FEMS Microbiol Lett 17:277-282
255. Work E, Griffiths H (1968) Morphology and chemistry of cell walls of *Micrococcus radiodurans.* J Bacteriol 95:641-657
256. Yamada H, Tsukagoshi N, Udaka S (1981) Morphological alterations of cell wall concomitant with protein release in a protein-producing bacterium, *Bacillus brevis* 47. J Bacteriol 148:322-332
257. Yoshii Z (1966) Macromolecular pattern in the cell wall of *Bacillus alvei.* Proc 6th Int Congr Electron Microsc, Kyoto, 2:277
258. Zabel H-P, König H, Winter J (1984) Isolation and characterization of a new coccoid methanogen, *Methanogenium tatii* spec. nov. from a solfataric field on Mount Tatio. Arch Microbiol 37:308-315
259. Zabel H-P, König H, Winter J (1985) Emended description of *Methanogenium thermophilicum,* Rivard and Smith, and assignment of new isolates to this species. System Appl Microbiol 6:72-78
260. Zehnder AJB, Huser BA, Brock TD, Wuhrmann K (1980) Characterization of an acetate-decarboxylating, non-hydrogen-oxidizing methane bacterium. Arch Microbiol 124:1-11
261. Zellner G, Alten C, Stackebrandt E, Conway de Macario E, Winter J (1987) Isolation and characterization of *Methanocorpusculum parvum,* gen. nov., spec. nov., a new tungsten requiring, coccoid methanogen. Arch Microbiol 147:13-20
262. Zillig W, Holz I, Janekovic D, Schäfer W, Reiter WD (1983) The archaebacterium *Thermococcus celer* represents a novel genus within the thermophilic branch of the archaebacteria. System Appl Microbiol 4:88-94
263. Zillig W, Stetter KO, Wunderl S, Schulz W, Priess H, Scholz I (1980) The *Sulfolobus-"Caldariella"* group: taxonomy on the basis of structure of DNA-dependent RNA polymerases. Arch Microbiol 125:259-269
264. Zillig W, Stetter KO, Schäfer W, Janekovic D, Wunderl S, Holz I, Palm P (1981) *Thermoproteales:* a novel type of extremely thermoacidophilic anaerobic archaebacteria isolated from icelandic solfataras. Zentralbl Bakteriol Hyg I Abt Orig C2:205-227
265. Zillig W, Gierl A, Schreiber G, Wunderl S, Janekovic D, Stetter KO, Klenk HP (1983) The archaebacterium *Thermofilum pendens* represents, a novel genus of the thermophilic, anaerobic sulfur respiring *Thermoproteales.* System Appl Microbiol 4:79-87
266. Zillig W, Stetter KO, Prangishvilli D, Schäfer W, Wunderl S, Janekovic D, Holz I, Palm P (1982) *Desulfurococcacáe,* the second family of the extremely thermophilic, anaerobic, sulfur-respiring *Thermoproteales.* Zentralbl Bakteriol Hyg I Abt Orig C3:304-317

Subject Index

MIX
Papier aus verantwortungsvollen Quellen
Paper from responsible sources
FSC® C105338

If you have any concerns about our products,
you can contact us on
ProductSafety@springernature.com

In case Publisher is established outside the EU,
the EU authorized representative is:
Springer Nature Customer Service Center GmbH
Europaplatz 3, 69115 Heidelberg, Germany

Printed by Libri Plureos GmbH
in Hamburg, Germany